Meeting the Balance of Electricity Supply and Demand in Latin America and the Caribbean

Meeting the Balance of Electricity Supply and Demand in Latin America and the Caribbean

Rigoberto Ariel Yépez-García, Todd M. Johnson, and Luis Alberto Andrés

THE WORLD BANK
Washington, D.C.

Cover photo: Gerencia de Communications Social, Comisión Federal de Electricidad; transmission lines in Cindad del Carmen, Campeche, Mexico.
Cover design: Naylor Design, Inc.

Contents

Foreword

Economic growth in the Latin America and the Caribbean Region has picked up considerably during the past decade or so. This growth has been aided by widespread (earlier and more recent) investments in power generation, transmission, and distribution that increased the provision of electricity services to households, commerce, and industry.

During the coming decades, the overarching goal of the power sector in the region will be to ensure an adequate supply of electricity for all consumers, produced in an efficient and clean manner and at the lowest possible cost. Given the region's more rapid economic growth and the currently tight supply-demand balance, the key challenge for the sector will be to meet the increasing demand for electricity without resorting to ever higher prices. But how fast will electricity demand grow? And are there ways to curb the growth in electricity demand without curtailing economic development? With even modest rates of economic growth, the demand for electricity is likely to double over the next 20 years. Under any scenario, the supply of electric power will need to expand. However, the *nature* of that expansion will have broad ramifications for the economies of the region in terms of cost, reliability, and environmental impact.

This report evaluates a number of critical issues for the power sector in the region in the coming two decades. These include the expected rates of increase in the demand for electricity, the required supply of new generating capacity, the technology and fuel mix of that generating capacity, and the carbon dioxide emissions of the sector. One of the key contributions of this study is the aggregation of individual country plans to the regional and subregional levels, using a consistent set of data and a common methodology. The report also assesses the important roles of hydropower and natural gas, the way other clean and low-carbon resources can be expanded, the potential and benefits of greater electricity trade, and the role of energy efficiency.

By considering the region as a whole, the report highlights the role that individual countries will play in shaping the region's aggregate power sector. Intended for policy makers and other experts, this volume makes a real contribution to a better understanding of the options for the power sector in Latin America and the Caribbean and of the implications of alternative development pathways for the region's economic growth and environment.

Augusto de la Torre
Chief Economist
Latin America and
  the Caribbean Region

Philippe Benoit
Sector Manager, Energy
Latin America and
  the Caribbean Region

Acknowledgments

This report is the product of a collaborative effort of three units of the Latin America and the Caribbean Region of the World Bank: the Energy Unit and the Economics Unit of the Sustainable Development Department and the Office of the Chief Economist. The task team leaders were Rigoberto Ariel Yépez-García, Todd M. Johnson (Latin America and the Caribbean Sustainable Development Energy Unit), and Luis Alberto Andrés (South Asia Sustainable Development). A special note of thanks goes to Luis Enrique Garcia (Latin America and the Caribbean Sustainable Development Energy Unit) for his assistance on the analytical work. Other World Bank staff members who contributed to the report include Pamela Sud, Ricardo Sierra, Aiga Stokenberga, Daniel Benitez, Tina Soreide, Govinda Timilsina, and Barbara Cunha. A number of consultants also provided important inputs for the report, including Roberto Gomelsky, Angela Cadena, Alberto Brugman, Marianna Iootty de Paiva Dias, and Ananda Covindassamy. The team specially thanks Alan Poole for his valuable contributions to the report and helpful suggestions. The team is grateful to the peer reviewers, Luiz Maurer and Marcelino Madrigal, for their valuable comments, as well as to John Nash and Juan Miguel Cayo. The report is the result of a concept note developed by Jordan Schwartz, Philippe Benoit, and Luis Alberto Andrés. For their overall guidance and detailed review of multiple drafts, the team would like to thank Philippe Benoit, Jordan Schwartz, Augusto de la Torre, and Francisco Ferreira.

The financial and technical support by the Energy Sector Management Assistance Program (ESMAP) is gratefully acknowledged. ESMAP is governed and funded by a Consultative Group composed of official bilateral donors and multilateral institutions, representing Australia, Austria, Canada, Denmark, Finland, France, Germany, Iceland, the Netherlands, Norway, Sweden, the United Kingdom, and the World Bank Group.

Abbreviations

AMN	Asociación Mercosur de Normalización (Mercosur Standards Organization
BIGTCC	biomass gasifier with combined-cycle gas turbines
CFE	Comisión Federal de Electricidad (México)
CIER	Comisión de Integración Energética Regional (Regional Energy Integration Commission)
CO_2	carbon dioxide
COPANT	Comisión Panamericana de Normas Técnicas (Pan-American Standards Commission)
CRIE	Comisión Regional de Interconexión Electrica
DSM	demand-side management
EE	energy efficiency
ENDESA	Empresa Nacional de Electricidad S.A.
EPR	Empresa Propietaria de la Red
ESCO	energy service company
FIDE	Fideicomiso para el Ahorro de Energía Eléctrica (Trust Fund for Electricity Savings)
GHG	greenhouse gas
GDP	gross domestic product
GW	gigawatt

GWh	gigawatt-hour
ICEPAC	Illustrative Country Expansion Plans Adjusted and Constrained
IMF	International Monetary Fund
INDE	Instituto Nacional de Electrificación (National Institute of Electrification)
ISA	Interconexión Eléctrica S.A. E.S.P.
kV	kilovolt
LAC	Latin America and the Caribbean Region
MER	Mercado Eléctrico Regional (Regional Electricity Market)
MME	Ministry of Mines and Energy (Guatemala)
MW	megawatt
OECD	Organisation for Economic Co-operation and Development
OLADE	Organización Latinoamericana de Energía (Latin American Energy Organization)
PECC	Programa Especial de Cambio Climático (Special Program for Climate Change)
PROINFA	Programa de Incentivo às Fontes Alternativas de Energía Eléctrica
RTR	Red de Transmisión Regional
SENER	Secretaría de Energía (Mexico)
SIEPAC	Sistema de Interconexión Eléctrica de los Países de América Central (Central American Electrical Interconnection System)
SUPER	Sistema Unificado de Planificación Eléctrica Regional
TWh	terawatt-hour

Executive Summary

Introduction

The development of the power sector will be critical for the Latin America and the Caribbean region's economic growth over the coming decades. Economic and social development in the region over the past 40 years have been supported by a widespread program of electrification that has greatly increased the provision of electricity services to households, commerce, and industry. Over the next 20 years, the supply of electric power will need to expand to meet the growing demand for electricity, but how the production and use of electricity develop will have broad ramifications for the diverse economies and societies of the region.

Among the key challenges for the development of the power sector in Latin America and the Caribbean over the next 20 years, which have implications for near-term investments and policies, are the following:

- **Electricity for growth and access.** Between now and 2030, how much electricity will be required by individual countries and the region as a whole to satisfy the needs of economic development and to provide access to those without electric power? How much investment—in new generating capacity, transmission, and distribution infrastructure—will be required?

- **Energy security.** Is the expansion of electricity supply likely to increase or decrease the diversity of the generation technology mix? Will this expansion make use of domestic or imported energy resources, and will the fuel sources be vulnerable to supply disruptions or large price shocks?
- **Energy efficiency.** Will the supply of electricity be provided at the least overall cost to national economies? Will the power generation technologies chosen be the most efficient, and are there opportunities to avoid new power generation capacity by producing and consuming electricity more efficiently?
- **Environmental sustainability.** What will be the trend in the role of natural gas, hydropower, and other clean and low-carbon electricity supplies in relation to petroleum and coal? What policies and regulatory regimes would help to promote low-carbon development in the power sector?
- **Regulatory framework.** What changes to the regulatory framework are needed to allow the power sector to meet increasing demand, address growing environmental concerns, and attract private capital to reduce the financial burden on government budgets?

The objective of this report is to provide an assessment of the electric power sector in Latin America and the Caribbean to 2030 and, in the process, provide initial answers to the previous questions. The report begins by examining the history of the power sector in the region, specifically the development of electricity production and the associated policies and institutions. Looking to the future, the report relies on the most recent and consistent regional data set from the Latin American Energy Organization (Organización Latinoamericana de Energía, or OLADE) and uses a common modeling framework to examine possible future trends in electricity supply. To the extent possible, the modeling framework attempts to reflect the current power expansion supply plans of the countries in the region. Critical outputs from the modeling analysis are presented in the form of a baseline scenario to 2030 for countries and subregions and include the demand for electricity, the total new supply of electric generating capacity needed, the technology and fuel mix of the generating capacity, and the carbon dioxide (CO_2) emissions of the sector. The report also examines a range of options and the policies needed to meet the future electricity supply challenges in the region. Among these options are the expansion of the use of hydropower, natural gas, and non-hydro renewable energy resources; increased regional electricity trade; and efficiency improvements on both the supply and the demand sides.

Historical Development of the Electric Power Sector in Latin America and the Caribbean

The power sector in Latin America and the Caribbean has experienced steady growth since the 1970s. Regional electricity production grew at an average rate of 5.9 percent per year between 1970 and 2005, compared with the worldwide average of 4.3 percent over the same period.

Six countries account for 84 percent of total electricity production in the Latin America and Caribbean region (figure ES.1). Brazil is the largest electricity producer (36 percent), followed by Mexico (21 percent), Argentina (9 percent), República Bolivariana de Venezuela (9 percent), Colombia (5 percent), and Chile (4 percent). Paraguay is a significant producer (5 percent) through its share of production from the gigantic Itaipu hydrostation; however, the majority of the electricity produced by Paraguay is sold to Brazil.

Figure ES.1   Market Share of Total Electricity Production, 2005

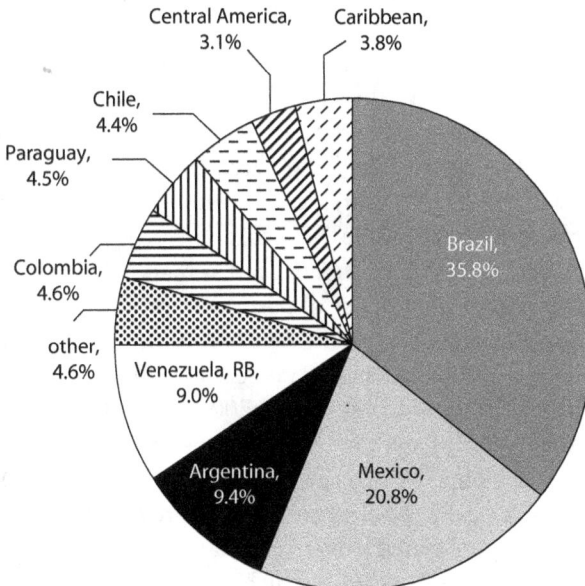

Source: Authors' elaboration based on World Development Indicators database 2009.

There are large disparities in electricity access rates both between and within countries. Despite the overall impression of affluence generated by the average growth rates for electricity production and consumption, countries in Latin America and the Caribbean face significant supply-demand imbalances (especially during dry years), and there are large differences in connection rates and affordability. For example, an estimated 34 million people in the region are without access to electricity (such as Peru, 6.5 million; Brazil, 4.3 million; Colombia, 3.0 million; and Guatemala, 2.7 million).

Hydroelectricity has been the dominant source of electricity for the region, but its share has been declining. Historically, hydroelectricity has provided the largest share of electricity in Latin America and the Caribbean, with the largest producer, Brazil, generating about 87 percent of its electricity from hydroelectricity in 2005. For the region as a whole, hydroelectricity provided 59 percent of electricity supply (2005), the highest share from hydroelectricity of any region in the world. Nonetheless, hydropower's share has been declining over the past decade (from 66 percent in 1995), and there are indications that the downward trend will continue.

Natural gas usage has been growing. A significant trend in the power sector in Latin America and the Caribbean over the past 15 years has been the growth in the use of natural gas—10 percent of generation capacity in 1995 rising to 19 percent in 2005 (over the same period, natural gas capacity rose from 15 percent to 38 percent in Mexico and 19 percent to 33 percent in the Southern Cone). The increase in natural gas has been the result of a variety of reasons, including the efficiency (and cleanliness) of natural gas for power generation and the increased production and trade of natural gas among countries of the region (Argentina, Bolivia, Brazil, Chile, Mexico, and Peru).

Petroleum use has declined overall but remains significant for some countries and subregions. The use of petroleum products (mainly fuel oil and diesel) for power generation has been significant for some subregions (75 percent in the Caribbean and 40 percent in Central America in 2005) and countries (31 percent in Mexico in 2005, down from 58 percent in 1985). For the region as a whole, however, the share of oil-fired generation accounted for only 14 percent of power generation in 2005, down from 20 percent in 1985. Dealing with the unpredictable fluctuations in oil prices and the associated effect on balance of payments remains a central concern for countries with a high share of oil in their electricity and overall energy-supply mixes.

Coal and other energy sources account for a small share of power generation in Latin America and the Caribbean. Coal use accounted for about 6 percent of power production in 2005, up from about 4 percent in 1985. The only country in the region with significant coal development plans is Colombia, which possesses the region's largest coal reserves. All other sources of power generation (including nuclear energy, wind, geothermal energy, and biomass) accounted for less than 2 percent of overall power generation in the region in 2005.

Electricity trade in Latin America and the Caribbean has been limited, but there is potential for growth with new interconnections. Trade has significant potential for balancing electricity supply and demand between countries and subregions, and the potential for increased trade has been facilitated by the construction of electricity transmission infrastructure, such as within Central America (Sistema de Interconexión Eléctrica de los Países de América Central, or SIEPAC) and between countries (Mexico and Guatemala; Colombia and Ecuador). However, with the exception of the sale of hydroelectricity from Paraguay to Brazil and Argentina, electricity trade in the region remains limited, both in absolute magnitude and as a percentage of overall demand.

From a global environmental perspective, the Latin America and the Caribbean region has the least carbon-intensive electricity sector of any region in the world, but carbon intensity has been rising. The low level of greenhouse gas emissions per unit of electricity production in the region is the result of the high share of hydroelectricity. However, the carbon-intensity of the power sector has been rising with the increasing share of fossil fuels (including natural gas) over the past decade, and this trend is expected to continue under a baseline scenario.

The regulatory framework for the power sector has experienced dramatic changes in the region. Beginning in the 1990s, new independent regulatory agencies were created, large state-owned companies were unbundled and privatized, and competitive market-oriented frameworks were implemented in a number of countries. However, the state remains an important player throughout the region in the power sector through the ownership of companies involved in generation, transmission, and distribution.

Baseline Electricity Supply Scenario

Modeling of electricity supply to 2030 was undertaken for the Latin America and the Caribbean region. For illustration of the implications of current trends in electricity development—for individual countries, sub-regions, and the region as a whole—scenarios of electricity supply to 2030 were created using a simple electricity demand function and a detailed energy-supply planning model:

- **Demand function.** The demand for electricity was estimated using gross domestic product (GDP) forecasts from the International Monetary Fund for each country to 2014 (IMF 2009). For the period 2015–30, a common set of economic assumptions—based on an average GDP growth rate of 3 percent per year—was used.

- **Supply model.** An electricity supply scenario—intended to reflect the current power sector expansion plans in the region—was created to illustrate electricity supply trends in the Latin America and the Caribbean region. Using OLADE's SUPER (Sistema Unificado de Planificación Eléctrica Regional) model and consistent country-level data, the authors created an electricity supply scenario that relies on the latest power sector plans of individual countries in the region and that satisfies the demand-function estimates.

- **ICEPAC (Illustrative Country Expansion Plans Adjusted and Constrained) scenario.** Using the demand estimates and the SUPER supply model, the authors created a baseline scenario for the next two decades. The "Illustrative" scenario is based on (a) "Country Expansion Plans" to 2030 (where available), which are then (b) "Adjusted" to account for missing data and to extrapolate country expansion plans (most of which are available to 2018 or 2020), and then (c) "Constrained" so as not to exceed energy resource potential (such as domestic hydroelectric resources),using a database of international technology supply costs that places a cost-minimizing constraint on the electricity supply model. From the ICEPAC scenario, it is possible to observe what would happen to the scale and structure of electricity supply, the financing that would be needed for new investment, and future CO_2 emissions from the power sector.

The key results of the electricity modeling exercise, reflecting current country expansion plans in the region, are the following:

- By 2030, with a modest rate of economic growth, the region's demand for electricity would reach nearly 2,500 terawatt-hours (TWh), up from about 1,150 TWh in 2008 (figure ES.2 and table ES.1). Electricity demand in Brazil would more than double to about 1,090 TWh. A total of 239 gigawatts (GW) of new electricity-generation capacity would be needed to match demand, with Brazil adding about 97 GW, the Southern Cone 45 GW, Mexico 44 GW, the Andean Zone 30 GW, Central America 15 GW, and the Caribbean 7 GW.

o Hydropower and natural gas would provide the majority of additional power capacity. Although the share of hydropower will continue to decline, the combined share of hydropower and natural gas

Figure ES.2   Regional Electricity Demand Scenario

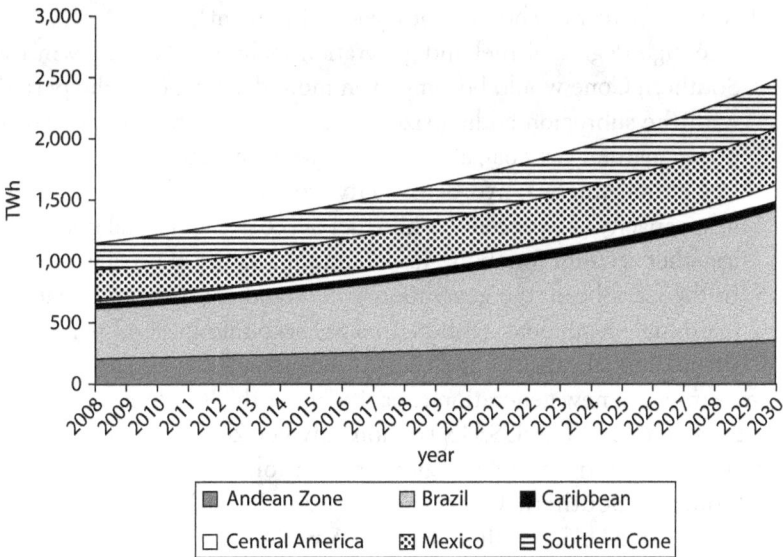

Source: Authors' elaboration based on optimization model.

Table ES.1   Regional Electricity Demand Scenario

Country or subregion	Average annual growth (%)
Andean Zone	2.8
Brazil	4.7
Caribbean	3.2
Central America	5.3
Mexico	3.4
Southern Cone	2.8
Average	**3.7**

Source: Authors' elaboration.

will be higher. There would continue to be a decline in the use of petroleum and a slight increase in nuclear (concentrated in Argentina) and non-hydro renewables.

o Despite the decline in hydropower's share, many countries and subregions are planning to substantially increase the absolute capacity of hydropower over the coming decades, including the Andean Zone, Brazil, Central America, and the Southern Cone. The aggregate increase in hydroelectric capacity by 2030 would be about 85 GW under the ICEPAC scenario.

o In Mexico, natural gas is estimated to be the most important fuel for new power generation (51 percent of new capacity), followed by coal (23 percent), hydroelectricity (14 percent), diesel (8 percent), wind (3 percent), and nuclear energy (1 percent).

o The high degree of fuel and generation technology diversity in the Southern Cone would become even more dynamic over the period, with the subregion adding sizeable generating capacity for hydropower, natural gas, coal, and nuclear energy in Argentina.

o In Central America, hydroelectricity would be the largest source of new capacity (45 percent), while fuel oil, coal, and natural gas would together account for about 45 percent of additional capacity.

o In the Caribbean, the generation mix would continue to be largely fossil fuel–dependent, with natural gas accounting for 43 percent of the additional capacity and coal 23 percent.

• Investment in new generation capacity under the ICEPAC scenarios is estimated to be about US$430 billion between 2008 and 2030. Investments by country and subregion are the following: Brazil US$182 billion, the Southern Cone and Mexico US$78 billion each, the Andean Zone US$58 billion, Central America US$25 billion, and the Caribbean US$9 billion.

• CO_2 emissions from electricity generation in Latin America and the Caribbean would more than double between 2008 and 2030 as a result of a decline in hydroelectricity and an increase in fossil fuels.

The modeling exercise demonstrates that with modest economic growth, there will be a need for a large expansion of power-generating capacity in the region, mainly fueled by hydropower and natural gas (figure ES.3 and table ES.2). Hydropower and natural gas are the least-cost sources of new power capacity and will contribute to both local and global environmental sustainability. To meet the optimistic goals for hydropower and natural gas—which are not explained in the modeling

Figure ES.3   Electricity Generation by Technology in Latin America and the Caribbean, ICEPAC Scenario

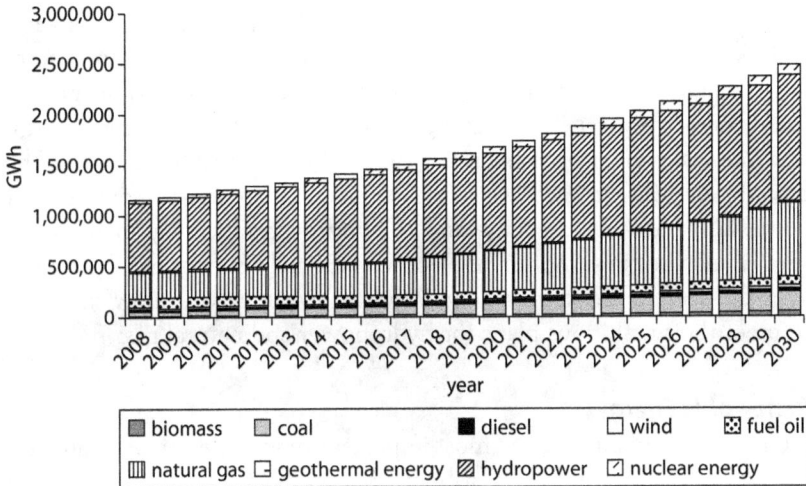

Source: Authors' elaboration based on optimization model.

Table ES.2   Electricity Generation by Technology in Latin America and the Caribbean, ICEPAC Scenario
percent

Source	Mix (2008)	Mix (2030)
Biomass	0.5	2.0
Coal	4.6	7.9
Diesel	2.3	1.2
Wind	0.1	1.3
Fuel oil	8.4	3.3
Natural gas	22.0	29.4
Geothermal energy	1.0	0.8
Hydropower	58.6	50.0
Nuclear energy	2.8	4.2

Source: Authors' elaboration based on optimization model.

analysis—many Latin American and Caribbean countries need to reform their respective regulatory, contracting, and licensing processes. There are also a number of other options to help the region meet its electric power needs that do not feature prominently in most national electricity expansion plans and, thus, are not captured in the modeling analysis. These options include an expansion of non-hydro renewables, greater regional electricity trade, and enhanced energy efficiency.

Options for Meeting Latin America and the Caribbean's Growing Electricity Needs

Despite the numerous strengths of the modeling exercise, it does not include a number of options that are becoming increasingly attractive to power planners in Latin America. Among these options are (a) the greater use of non-hydro renewable energy, including wind, geothermal energy, and biomass, which have been growing in importance globally over the past decade; (b) an increasing role of electricity trade to complement domestic generating capacity; and (c) improved efficiency in both the supply and the consumption of electricity. Equally important, if not more so, is the need for policies and regulatory reforms that will allow countries to meet their ambitious plans for hydropower and natural gas.

Renewable Energy

Hydroelectricity is by far the most important renewable energy source for the Latin America and the Caribbean region, both historically and over the coming two decades as indicated in the country power-expansion plans. As demonstrated in the modeling exercise, even with a dramatic expansion, the share of hydroelectricity in total electricity generation is likely to decline. If the region is to maintain the current proportion of about 60 percent of renewable electricity in its generation mix, the use of non-hydro renewable energy would need to expand by about 150 TWh by 2030 (with non-hydro renewables increasing from 2 percent to 4 percent of total power generation), while still meeting the aggressive targets for hydropower.

Hydropower. For the region to maintain the current high share of hydropower, it is necessary to develop hydropower resources in those countries—Colombia, Ecuador, and Peru—that possess more than half of the hydropower potential outside of Brazil, and that today have developed only 10 percent of their hydroelectric potential. Greater integration among regional power markets could help to justify and attract financing for the development of larger hydropower projects in these countries.

Other low-carbon options. In addition to hydropower, a number of other promising low-carbon options for electricity production in Latin America and the Caribbean could allow the region to maintain its high share of renewables, even under a low-hydroelectricity scenario. Under such a scenario, about 480 TWh of new output from non-hydro sources would be required, which could potentially be supplied through a combination of electricity production from wind (220–340 TWh), biomass (55–150 TWh), and geothermal energy (25–125 TWh). Other renewable power-generation technologies were not evaluated in this report.[1]

Electricity Trade

Regional electricity trade could help meet Latin America's electricity needs by making better use of regional energy resources (such as hydro-power that typically has scale economies) and by linking a larger set of generators and consumers in a single market. The potential for trade is being facilitated by new interconnections, including those between Mexico and Central America, between countries of Central America (through the SIEPAC system), between Colombia and Panama, and between countries of South America.[2] The history of energy trade in the region provides valuable lessons, on both the benefits and the constraints, to greater regional integration of electricity markets.

Electricity trade has a number of potential benefits compared with exclusive reliance on domestic generation. Trade can (a) enhance the reliability of the local network by linking together a larger number of generation sources and, thus, increasing the diversity and competitive-ness of generation; (b) have a positive effect on reducing capital invest-ment and generation costs (both operational and capital expenses) because of the economies of scale associated with power generation from large facilities and the reduction in the need for reserve capacity; (c) lead to an important reduction of recurrent expenses because countries do not need to import costly fuels; (d) free up capital from domestic elec-tricity capacity expansion programs; and (e) permit the linking of areas with different hydrology or wind regimes, thus increasing the supply of "firm" energy from variable or intermittent energy sources such as hydro-power and wind.

According to a quantitative exercise undertaken for Central America, increased trade could increase the share of hydroelectricity from 46 percent to 54 percent, simply by relying on hydroelectricity plants that could be built in Central America. Tapping increased hydroelectricity from North American or South American markets would likely raise the share of hydro-power in the subregion. The increase in hydroelectricity (from Central America only) and the consequent reduction in necessary thermal power would reduce CO_2 emissions by 14 percent. There also would be significant savings from trade in domestic investment in the power sector by lowering the need for reserve capacity.

New institutions and regulations are needed to facilitate electricity trade. Despite the potential benefits of trade, experience shows that countries in Latin America and the Caribbean have not taken advantage of electricity trade for a number of reasons, including perceived energy security and national sovereignty issues. Interconnection, such as the SIEPAC system in Central America, is an essential step in the process, but a regulatory framework to facilitate trade between different countries with different regulatory policies and power sector institutions is also needed.

Energy Efficiency

Although there have been no comprehensive studies of energy efficiency potential in Latin America and the Caribbean—and this report does not attempt to fill that void—there is sufficient evidence to show significant untapped energy-efficiency potential in the region. In addition, according to the energy-efficiency and conservation programs that have been implemented in the region, efficiency is one of the least-cost ways of satisfying growing energy demand.

Energy-efficiency gains can be achieved on the supply side by improving the production of electricity and by reducing transmission and distribution losses. Electricity distribution losses alone in the region in 2005 were equal to the entire electricity consumption of Argentina, Chile, and Colombia combined. Distribution losses vary significantly in the region, ranging from a low of 6 percent in Chile to a high of more than 40 percent in the Dominican Republic, with a Latin American and Caribbean average of about 16 percent. If distribution losses could be reduced to the levels of the best performers in the region over the next 20 years, annual electricity savings from distribution improvements alone could reduce demand by about 78 TWh (6 percent of the incremental demand of 1,325 TWh) by 2030.

On the demand side, efficiency can be improved by adopting policies and programs that encourage the efficient consumption of electricity by end users. Among the energy-efficiency measures that can be expanded in the region are standards for widely used industrial and residential equipment; building codes; consumer education; and energy management programs within industry, the buildings sector, and public utilities. Electric motors, pumps, fans, and compressors, which are estimated to account for as much as two-thirds of industrial electricity consumption worldwide, can reduce their electricity consumption by about 40 percent through the use of variable-speed drives. The Inter-American Development

Bank has estimated that electricity consumption in Latin America and the Caribbean could be reduced by about 10 percent (143 TWh) over the coming decade through investment in widely available energy-efficiency equipment and technologies, and that these savings could be achieved at about one-third the cost of installing new generation capacity.[3] Other estimates of the potential for demand-side efficiency improvements, based on energy-efficiency programs implemented in the region, range from 18 to 30 percent of estimated additional electricity demand in the region by 2030.[4]

Additional incentives—such as electricity tariff and subsidy reform—could improve the efficiency of electricity use. Although industry often has a sufficient direct financial incentive to improve its electricity efficiency—depending in part on the level and structure of electricity tariffs—the market alone often provides inadequate incentives to promote energy efficiency in the residential, buildings, and public sectors. Overcoming information, principal-agent, budgeting and finance, and regulatory constraints through dedicated public energy-efficiency programs can help improve efficiency in these sectors.

Summary and Conclusions

Under modest GDP growth assumptions, the demand for electricity in Latin America and the Caribbean would more than double by 2030. Under current expansion plans, the region would need to add more than 239 GW of new power-generating capacity to meet demand. A higher rate of economic growth, or a higher demand for electricity, would require even more new capacity. Under any economic scenario, it will be challenging for the region to meet future electricity demand by relying on current plans for power sector expansion.

Under the baseline scenario, the vast majority of the increase in generating capacity between now and 2030 would be met by hydropower (36 percent) and natural gas (35 percent). The baseline scenario represents a "best case" scenario, because many of the country expansion plans for hydropower and natural gas are already quite optimistic. Under the baseline, an estimated 85 GW of new hydro capacity would be required, compared with only 76 GW built in the region over the past 20 years. In addition, in some countries many of the best sites—in terms of construction costs and low environmental and social impacts—have already been developed. The relatively long payback periods, high capital costs, and environmental and social risks have reduced private sector involvement in hydroelectric plants and, thus, reduced the scale and pace of hydro development.

Natural gas is one of the Latin America and the Caribbean region's best alternatives (both economically and environmentally) for new power-generating capacity, and under the baseline, gas-fired capacity would grow from 60 GW to more than 144 GW in 2030. Many countries in the region have been expanding the use of natural gas for power generation using efficient combined-cycle technology. However, in some countries of the region low "preferential" prices for natural gas and low electricity tariffs have resulted in the inefficient use of natural gas, including the use of "open-cycle" gas plants, as well as reduced incentives for producing and distributing gas for power generation.

Alternatives exist for meeting future electricity needs. The analysis suggests that meeting the demand for electricity in Latin America and the Caribbean can be achieved not only by building new generating capacity, but also by relying on an increased supply of non-hydro renewables, by expanding electricity trade, and by making use of supply- and demand-side energy efficiency to lower the overall demand for electricity.

- **Non-hydro renewables.** There is significant potential for expanding the use of non-hydro renewables in the region, including extensive wind resources from Mexico to Argentina, geothermal resources along the tectonically active Pacific rim and in the Caribbean, and biomass resources (such as sugarcane bagasse) throughout the region. These energy resources can help diversify the overall electricity supply mix in Latin America and the Caribbean, and in many instances, non-hydro renewable technologies are becoming cost-competitive with conventional power technologies.

- **Increased electricity trade.** Trade could provide significant new capacity by enlarging the region's electricity market and lowering overall supply costs in the process. Increased trade could also help the region make use of its hydroelectric and other energy resources by linking energy supplies to a larger market, thus justifying some larger-scale projects and attracting regional investment.

- **Improved energy efficiency.** Energy efficiency is the most cost-effective way to meet future energy demand, with significant potential on both the supply and the demand sides. Many investments pay for themselves quickly, such as reducing transmission and distribution losses and tapping the huge amount of cogeneration potential in industry. The potential for improving the efficiency of energy use is even greater on the consumption side, ranging from residential and commercial lighting to broadly used electrical appliances (such as refrigerators and air conditioners) to industrial motors and pumps. Recent studies in Brazil and Mexico confirm the extent of the energy-efficiency potential that could be tapped at low cost.

The aggregate effect of these alternatives—in terms of lowering the requirements for new generation capacity, much of it thermal—could be large. The analysis suggests the following: (a) an aggressive program to expand non-hydro renewables could provide between 15 and 30 percent of the total electricity supply by 2030; (b) increased trade could lower electricity costs by allowing the development of larger-scale and, in some cases, regional projects, including more renewables and also could reduce investments in reserve capacity; and (3) overall electricity demand in the region could be lowered by at least 10–15 percent through limited supply-side and demand-side energy-efficiency measures at a fraction of the cost of constructing new power generating capacity (figure ES.4).

Figure ES.4   Electricity Supply Mix in Latin America and the Caribbean, Various ICEPAC Scenarios

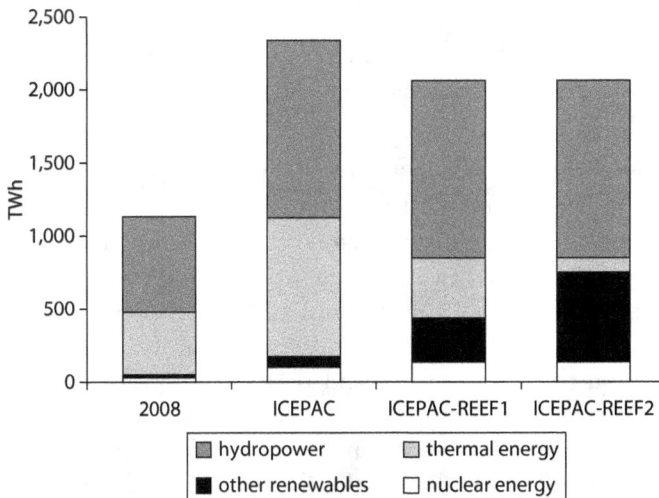

Source: Authors' elaboration based on optimization model.
Note: ICEPAC = Illustrative Country Expansion Plans Adjusted and Constrained, REEF = Renewable Energy and Energy Efficiency alternative scenarios.

Recommendations

There are a number of recommendations that flow from the conclusions of this report and that have been identified in other recent energy sector analyses by the World Bank.[5]

Strengthening Regulations and Market Design of Hydropower and Natural Gas Generation Projects

Hydropower and natural gas are indigenous and proven energy resources that can help the Latin America and the Caribbean region achieve a supply and demand balance for electricity over the next two decades. Meeting the proposed increases in hydroelectric capacity will require significant changes in the way power plants have been financed—a greater role for the public sector in regulating and guaranteeing hydroelectricity construction and a greater role for the private sector in taking on long-term construction or operation contracts. Improvement in the management of social and environmental issues is needed, and the licensing and commissioning process will need to be strengthened and streamlined. Reaching the hydroelectric goals for the region will require reforms in the way that hydroelectric plants are designed, prepared, and financed. Among the most important issues for natural gas development is gas pricing. Although low prices of domestic natural gas for power generation have been meant to stimulate natural gas development, they have resulted in the inefficient use of natural gas and a lack of incentives for its development. To ensure that regional natural gas resources are not wasted, a combination of pricing reforms and technology standards is required.

Supportive Policies for Renewable Energy and Energy Efficiency

Effective regulations and institutions are needed to provide incentives to both the public and the private sectors to invest in and develop renewable energy technologies and to promote energy-efficiency measures. A number of countries in Latin America and the Caribbean have put in place new renewable-energy laws and regulations, including tax credits, long-term purchasing contracts, and dispatch priorities for renewable energy. Countries also need payments that reflect marginal costs for the system, plus capacity payments, and payments that reflect the environmental benefits (both local and global) that renewable energy sources typically bring. Supportive policies for energy efficiency include standards for efficient plants and energy-consuming equipment, the establishment of utility and other programs for promoting and disseminating energy-efficient measures, and electricity tariffs that provide incentives for end users to acquire and use energy-efficient equipment and processes.

Domestic Energy Planning

The region needs to expand and strengthen its power sector planning. Although most countries carry out power sector planning, several have not yet developed electricity-specific supply-and-demand growth scenarios. In several country cases, the time horizon for planning is too short (12 years or less) or the plans are not updated frequently. Given the long-term nature of power sector investments, governments should engage in longer-horizon planning. Consultation with constituencies about medium- to long-term power sector development should be undertaken, should be realistic, and should include a diverse range of supply- and demand-management options.

Regional Power Sector Tools

This report made clear that there is a lack of tools for regional power sector analysis. There is thus a need for robust and user-friendly regional power-planning tools that can be used and discussed by individual countries, regional and international organizations, and the private sector. A regional planning tool should be able to optimize electricity-generation decisions not only for specific countries, but also across larger geographic regions. Furthermore, additional research is required to include demand price sensitivity, as well as to test the robustness of supply models with additional scenarios that do not limit the range of generation technologies or other measures (such as trade and efficiency) for meeting energy demand.

Reliable Inventory Information

One of the requirements for electricity development, especially for renewable-energy resources, is improved information on the size, quality, and location of energy resources. Publically supported inventories of wind and geothermal resources, for example, can help to reduce production risks and accelerate development of wind and geothermal resources.

Notes

1. Solar photovoltaic systems can play an important role in providing least-cost electricity access in isolated areas, but the total potential (in TWh) of both off-grid and on-grid photovoltaic systems is not expected to be large by 2030. This excludes the possibility that some countries may choose to support grid-connected or "rooftop" solar photovoltaic programs as is popular in several

Organisation for Economic Co-operation and Development countries (Germany, Japan, and the United States). Solar collectors for the production of hot water could provide a large amount of energy to the region, by substituting for residential and commercial hot water that is currently produced from electricity and natural gas. In Mexico, it is estimated that a large-scale program could displace as much as 23 TWh per year (Johnson and others 2010).

2. Currently, interconnections are between Brazil, Paraguay, Argentina, and Uruguay; between Mexico and Central America; and between Colombia, Ecuador, and República Bolivariana de Venezuela.

3. For reference,US$16 billion for 143 TWh (IDB 2009) equals US$112 million per TWh for energy efficiency, compared with an investment cost of US$430 billion for 1,325 TWh from the supply model in this report, which equals US$315 million per TWh.

4. For reference, 18 percent is the estimate for regional potential based on an extrapolation of Argentina's estimates of energy-efficiency potential in the industrial, residential, and commercial sectors in 2008.

5. Among the recent reports by the World Bank that address key policy and institutional issues of the power sector in Latin America are *Low-Carbon Development for Mexico* (Johnson and others 2010); *Brazil Low-carbon Country Case Study* (World Bank 2010a); *Peru: Downstream Natural Gas Study* (World Bank, forthcoming); "Peru: Overcoming the Barriers to Hydropower," (World Bank 2010b); *An Overview on Efficient Practices in Electricity Auctions* (World Bank, forthcoming).

References

IDB (International Development Bank). 2009. *Cómo economizar US$36.000 millones en electricidad (sin apagar las luces)*. Washington, DC: IDB.

IMF (International Monetary Fund). 2009. *World Economic Outlook, April 2009: Crisis and Recovery*. Washington, DC: IMF.

Johnson, T. M., C. Alatorre, Z. Romo, and F. Liu. 2010. *Low-Carbon Development for Mexico*. Washington, DC: World Bank.

World Bank. 2010a. *Brazil Low-carbon Country Case Study*. Washington, DC: World Bank.

———. 2010b. "Peru: Overcoming the Barriers to Hydropower." ESMAP Report 53719-PE, Energy Sector Management Assistance Program, Latin America and the Caribbean Region, World Bank, Washington, DC.

———. Forthcoming. *An Overview on Efficient Practices in Electricity Auctions*. Washington, DC: World Bank

———. Forthcoming. *Peru: Downstream Natural Gas Study*. Washington, DC: World Bank.

CHAPTER 1

Introduction

In Latin America and the Caribbean, economic and social development over the past 40 years has been supported by a widespread and largely successful program of electrification. That effort has greatly increased the provision of electricity services to households, commerce, and industry and has led to electricity access rates[1] that are among the highest in the developing world. Over the coming decades, the supply of electric power will need to expand to meet the growing demand for electricity, but how the production and use of electricity develops—in terms of the amount of new capacity, the technology mix, the source and type of fuel, the structure of demand, and the efficiency of consumption—will have broad economic, social, and environmental implications for Latin America and the Caribbean and the world.

A number of critical issues will affect the expansion of the power sector in Latin America and the Caribbean and, in turn, will have implications for that expansion. Among these issues are the rate of economic growth, the energy resources available in the region or through trade, the types of power technologies that are adopted, and the cost of power sector investments and the sources of financing. Government policies will also affect power sector expansion, including (a) distributional policies related to electricity access; (b) energy pricing and other policies affecting both demand and supply decisions; and (c) environmental policies at both the national and the international levels that will affect the technology and fuel mix of new generating capacity and the consumption decisions of households, government, and industry.

The objective of this report is to assess some of the key electricity challenges that Latin America and the Caribbean will face in the coming decades in meeting its development, security, efficiency, and environmental goals. The two focal areas of the report are (a) evaluation of the trends in the power supply mix and the implications of the generation mix on investments and environmental outcomes and (b) exploration of the region's options for making greater use of renewable and low-carbon energy resources, tapping the potential benefits of increased electricity trade, and mitigating the need for new capacity additions through energy efficiency improvements.

Electricity and Development

A sufficient supply of affordable and reliable electricity is a core precondition for the economic growth of Latin America and the Caribbean as well as for an improved quality of life for its poorest people. Industrial and commercial activity and the development of modern cities require electricity to power a broad range of end uses such as pumps and motors, HVAC (heating, ventilation, and air conditioning), municipal lighting, elevators, metro systems, and traffic signals. As a macroeconomic variable, electricity consumption is universally highly correlated with national income. Although the causality works in both directions—that is, electricity determines gross domestic product (GDP) and GDP determines electricity consumption—there is little doubt that electricity is a critical enabling factor for economic development.

Electricity and other modern forms of energy are also critical elements for improving the welfare of low-income groups as well as for allowing rural areas to develop economically in the broader sense, including providing opportunities for financial livelihoods, improving human health, and raising education levels. In addition to clean water, the provision of electricity is perhaps the most important way to reduce rural poverty and the reason rural electrification programs are often high on a country's national rural development agenda.

The social and economic benefits of providing electricity to households have been studied for more than three decades. One of the first and universal uses of electricity by households is for lighting, which allows activities (such as reading, studying, or doing household chores) to continue into the evening (Barnes, Peskin, and Fitzgerald 2003). Other household uses of electricity include for consumption purposes (radio, television, and cell phone charging) and productive uses (sewing machines, small appliances, and shop lighting), which can have important economic, social, and cultural benefits (World Bank 2004b). Women and children are usually the prime beneficiaries of electrification. A study in India found that women from homes with electricity were better able to manage paid work, household chores, and leisure time than women from homes without electricity (World Bank 2004a). Over the long term, evidence shows a positive relationship between electricity consumption and household income. Providing electricity to households can thus be seen as an overall positive investment for the economy.

Latin America and the Caribbean's Electricity Challenge

Despite the pressing need for reliable and affordable electricity, the power sector is characterized by a long-term planning and investment horizon, which is complicated by a number of risks and uncertainties. Among these are fluctuations in long-term demand; multiple sources of fuel and technologies for power generation; high and volatile prices of fossil fuels; political risks of relying on bilateral trade in fuels or electricity; and environmental and social effects associated with the production, transmission, and distribution of electricity. One certainty is that as countries in Latin America and the Caribbean become more prosperous, the demand for electricity will increase and the challenges of the region's economies to meet their energy requirements will intensify.

Among the electricity challenges that Latin America and the Caribbean (and other regions of the world) confront are the following:

- *Economic growth and access:* the need to expand the current electricity supply over the next 20 years to support increases in income and to provide electricity access to unelectrified households and communities
- *Energy security:* the increasing risk of electricity supply disruptions and price shocks as a result of the growing dependence on imported fossil fuels, and the constraints faced in tapping national and regional renewable (such as hydroelectricity and wind) and low-carbon (such as natural gas) energy resources
- *Economic efficiency:* the need to limit the cost of providing new electricity by promoting competitive contracts and financing by the private

sector, to improve the efficiency of supply, and to avoid new construction when and where demand-side efficiency is least-cost
- *Environmental sustainability:* the desire to incorporate environmental goals, both local and global, into power sector planning, policy making, and investment
- *Conducive Regulatory Framework:* the need to put in place policies and regulations that allow the power sector to meet increasing demand, to address growing environmental concerns, and to attract private capital to reduce the financial burden on government budgets.

Countries seek practical and feasible strategies to meet the increasing demands for power. Currently, a number of Latin American and Caribbean countries are experiencing inadequate supply, which has been a factor behind recent unplanned power outages. Demand has been increasing over the past decade at a rate of 5 percent per year on average, and although the recent financial turmoil is anticipated to result in lower levels of growth in the near term (and demand projections are being adjusted downward), over the medium to longer term, demand is anticipated to rebound, which will require additional investment. Regardless of the ultimate growth in electricity demand, putting in place rational short- and medium-term pricing structures and demand-side efficiency measures can help improve the ability to supply electricity.

In meeting future electricity demand, policy makers must consider how, when, and through what means they plan to scale up capital investment in the electricity sector. As mentioned earlier, the electricity supply and demand balance is tenuous in some Latin American and Caribbean countries, in both the short and the longer term. In today's environment of lower reserve margins, an underestimation of electricity demand or underinvestment in power supply can lead to brownouts or blackouts. Short-term imbalances can result from unanticipated demand (such as a hot summer or cold winter) or supply disruptions (such as drought) and are exacerbated by low reserve margins. Long-term investment decisions affect not only reserve margins, but also the security of supply, depending, for example, on the type and price volatility of the fuel.

It is important to identify trends in the Latin American and Caribbean electricity markets. An analysis of electricity demand and supply is an important input for decision making, allowing policy makers to explore the effect of different assumptions on future investments, such as the rate of growth of the economy, generation fuel prices, and environmental constraints. Electricity sector planning and policy making in Latin America and the Caribbean will thus require an analysis of (a) the future supply of power generation, including an identification of sector investment needs; (b) the mix of technologies from the perspective of cost, environmental impact, and diversification, among other factors; and (c) the opportunities for efficiency gains, notably in generation, distribution, and consumption. Such an analysis also allows policy makers to consider the effect of different policies on electricity demand and supply, such as tariffs and energy pricing, trade policies, efficiency norms and standards, and environmental regulations.

Like other large-scale infrastructure investments, power sector planning requires a long-term perspective. Even if future electricity demand could be well anticipated, it takes many years to plan and build new power generating capacity, transmission lines, and distribution networks. Thus, a modeling framework of at least a 20-year period is needed to see the effects of changes in the supply mix and to allow policy makers and power sector planners time to adjust their long-term expansion plans.

Although Latin American and Caribbean countries are relatively "clean" in terms of their electricity generation mix, they need to examine the effect of their long-term power expansion plans on carbon dioxide (CO_2) emissions. Even if one assumes that the majority of these countries will not have binding CO_2 emission reduction targets anytime in the near future, the carbon intensity of the power sector is important for a number of reasons. Knowing the trends in carbon intensity and the options for reducing them will be important for setting a country's baseline emissions trajectory, which will inevitably be negotiated with bilateral countries (such as Japan, those in Europe, or others with climate change mitigation legislation) for the selling of carbon offsets or with the international community (such as under the the Kyoto Protocol's Clean Development Mechanism or a subsequent system) for emission reduction credits. Understanding the trend in the power sector is also important for those countries (such as Mexico) that have made voluntary commitments to reduce their emissions. There are likely to be other incentive programs (such as the Global Environment Facility, the Clean Investment Funds, and new mechanisms under the United Nations Framework Convention on Climate Change) for countries that promote low-carbon development of their power sectors.

Scope of the Report and Methodology

In light of the electricity challenges outlined above, this report focuses on two main issues that have major implications for planning and investment in the electric power sector: (a) the future supply and demand for electricity in Latin America and the Caribbean, and (b) new options for meeting future electricity demand in light of international technology, market, and regulatory trends.

The first focus of the report is an analysis of future electricity demand and supply. Making use of the most complete regional data compiled by the regional electricity organization, the Latin American Energy Organization (Organización Latinoamericana de Energía, or OLADE),[2] and constructing a modeling framework, the report evaluates potential future trends in electricity supply and demand in Latin America and the Caribbean through 2030. Signifying that the estimates are not forecasts or predictions, the report uses the term *scenario* to reflect what would happen if certain policies and trends are pursued. In the case of electricity demand, a conservative estimate of economic growth has been used. In the case of electricity supply, the scenario developed is meant to reflect what would happen if countries continue along the path of their current long-run power expansion plans.

Although many Latin American and Caribbean countries have done detailed analyses of their own electricity demand and supply situations, they are less familiar with other countries' actions and even less aware of the way an individual country contributes to regional demand and supply. Therefore, this report provides the aggregation of individual country plans to the regional and subregional level using a consistent set of data and a common methodology.

Through the demand-and-supply scenario exercise, it is possible to view the implications for the role of different generation technologies and fuels in the electricity supply mix, to estimate the level of investment that would be required, and to view the changing carbon intensity of the power sector in Latin America and the Caribbean. Sensitivity analysis on the effect of a carbon tax (US\$20–US\$50 per ton of CO_2) on fossil fuel generating capacity is also undertaken.

The supply analysis makes use of the individual country expansion plans for the short- and medium-term that are available from national energy planning agencies. Country estimates are aggregated and presented in terms of six subregions: Mexico; Central America; the Caribbean; Andean countries (Bolivia, Colombia, Ecuador, Peru, and República Bolivariana de Venezuela); the Southern Cone (Argentina, Chile, Paraguay, and Uruguay); and Brazil.

The scenario analysis in chapter 3 applies an approach developed in partnership with OLADE. The study team estimated electricity demand using a simple regression model, with electricity demand becoming an input to OLADE's SUPER (Sistema Unificado de Planificación Eléctrica Regional) model. The SUPER model allows an evaluation of a diverse array of alternatives for expanding electricity generation in Latin America (details of the model are provided in chapter 3 and in appendix C).

The second focus of the report is the range of new or additional options for meeting electricity demand that are not explicitly or consistently considered within the national power sector plans of individual countries. The options discussed were selected on the basis of (a) international trends in power sector technology and market development and (b) environmental policies and regulations, particularly those related to climate change. The options include the potential for expanding non-hydro renewable sources of electricity generation, increased regional trade in electricity, and greater energy efficiency in both supply and demand of electricity.

The information on renewable energy, trade, and energy efficiency was provided through specific assessments commissioned for the study, drawing on expert opinion and analysis and making use of the World Bank's work in these topics, both regionally and globally. Like the modeling exercise, the options analysis presented in chapter 4 uses a timeframe of 2030 to allow a comparison with the potential contributions from non-hydro renewable energy, regional electricity trade, and energy efficiency.

Structure and Content of the Report

The report is organized as follows. Chapter 2 provides the historical context of the electric power sector in Latin America and the Caribbean by presenting past trends in that sector and providing global comparisons. The chapter examines historical trends in the production of electricity, by country and subregions, as well as the changes in the generation mix. The chapter also looks briefly at the trends in regional trade in electricity as well as the organization and regulation of the power sector in Latin America and the Caribbean. Given the recent international financial crisis, the chapter briefly discusses the way this situation has affected the power sector in the short term and the implications for the future.

Chapter 3 presents the results of the electricity modeling exercise that was conducted from present to 2030. The methodology, major assumptions, and strengths and weaknesses of the demand function and supply model are discussed, with further details about the models included in appendixes B and C. The chapter highlights the focus of the modeling analysis on the supply side and describes how the methodology reflects the current country expansion plans of the individual countries in Latin America and the Caribbean. The chapter then presents the results of the electricity scenario analysis, identifying regional and subregional patterns and discussing some of the key challenges posed by the results. The analysis identifies the amount of new power production capacity that would be needed under the baseline scenario, indicates the way the expansion would affect the generation mix, and concludes with the implications for the carbon-intensity of the power sector in Latin America and the Caribbean.

In light of the challenges posed by the modeling results from chapter 3, chapter 4 discusses other options for meeting the growing electricity needs of Latin America and the Caribbean that were not well addressed in the modeling exercise. Chapter 4 attempts to compensate for some of the inherent deficiencies of current planning by explicitly examining a number of important options for meeting future electricity demand. First, it presents an analysis of Latin American and Caribbean renewable energy potential, both hydro and non-hydro alternatives, and discusses the way that potential compares to the planned investments included in individual country expansion plans. Second, it examines the benefits and implications of regional electricity trade, explores efforts to promote cross-border integration, and discusses the range of obstacles to greater cross-border initiatives in Latin American and Caribbean. Third, it examines the role of supply-side and demand-side energy efficiency options, including estimates of the magnitude of potential efficiency gains.

Chapter 5 briefly discusses the overall conclusions of the study, including a comparative summary of the scenario assessment presented in chapter 3 and the options analysis presented in chapter 4. The chapter concludes with a set of recommendations related to energy policies, power sector regulations and institutions, and the benefits and requirements of achieving better long-term power sector planning.

Notes

1. The electricity access rate refers to the percentage of the population, sometimes measured as the percentage of households, that have access to electricity, either through a centralized electricity grid or through stand-alone household or community systems.

2. OLADE is the regional organization for Latin America and the Caribbean by which its member states undertake common efforts to achieve integration

and development of the energy market. As part of its coordinating and planning activities, OLADE developed a statistical database in the energy sector that includes information from member states on electricity and other energy sector historical series since 1970.

References

Barnes, Douglas, Henry Peskin, and Kevin Fitzgerald. 2003. "The Benefits of Rural Electrification in India: Implications for Education, Household Lighting, and Irrigation." Draft paper prepared for South Asia Energy and Infrastructure Unit, World Bank, Washington, DC.

World Bank. 2004a. "The Impact of Energy on Women's Lives in Rural India." ESMAP Report 276/04, Energy Sector Management Assistance Program, World Bank, Washington, DC.

———. 2004b. "Power Sector Reform and Rural Poor in Central America." ESMAP Report 297/05, Energy Sector Management Assistance Program, World Bank, Washington, DC.

CHAPTER 2

Historical Trends in the Electricity Sector

Before one considers where the electricity sector of Latin American and the Caribbean is headed, it is important to know where the sector has been. This chapter examines the development of the electric power sector in Latin America and the Caribbean since the 1970s, looking at the growth in production by countries and subregions, the changing mix of generation technologies, and the role that electricity trade has played in meeting electricity needs. The chapter also looks at the governance and regulatory structure for the power sector in different countries, which has implications for sector development, both past and future. Finally, the chapter discusses the effect of the global financial crisis, beginning in 2007–08, on both power demand and power supply. Before one examines the history of the power sector in Latin America and the Caribbean, it is useful to place the region in an international context.

Latin America and the Caribbean in a Global Context

Worldwide electricity production increased fourfold from 1970 to 2005, implying an average annual growth rate of 4.3 percent. In Latin America and the Caribbean, electricity production increased at a faster rate, growing by 5.9 percent over the same period. However, the growth rate over the period was uneven. During the 1970s, electricity production grew at an average rate of 8.7 percent per year, peaking in 1978–1979. In the 1980s and 1990s, electricity production grew by 5.4 percent and 4.3 percent, respectively. In 2001, there was an absolute decline in electricity production, the only year over the past 40 years in which production in the region decreased.

Figure 2.1 shows the global development of electricity production from 1985 to 2005. North America has been the largest electricity producer since the mid-1980s. Although Europe and Central Asia followed directly behind North America until the late 1990s, East Asia and the Pacific, led by China, became the second-largest producer in 2000. The East Asia and Pacific region historically has had higher annual growth rates than other regions, driving the overall growth in the developing world's electricity production as shown in figure 2.1. However, Latin America and the Caribbean has remained an important producer, ahead of South Asia, the Middle East and North Africa, and Sub-Saharan Africa.

Figure 2.1   Electricity Production

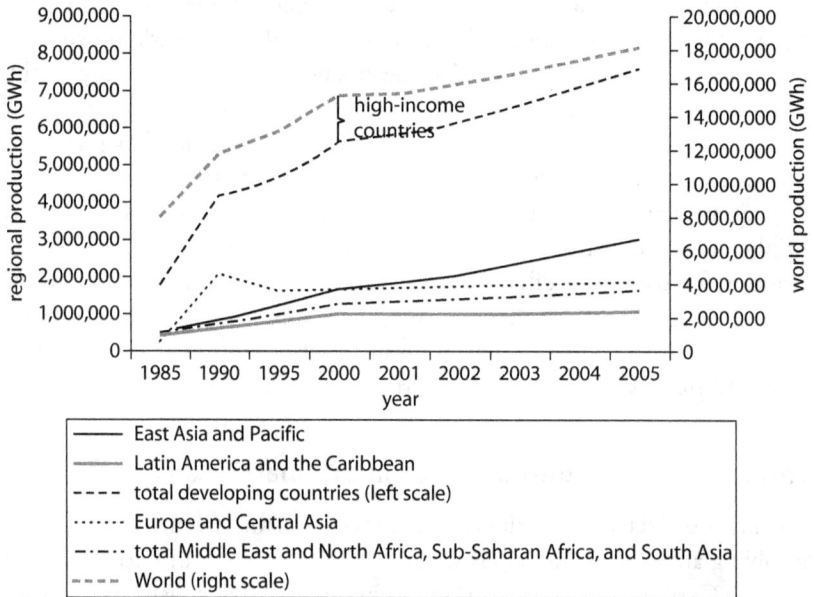

Source: Authors' elaboration based on World Development Indicators database 2009.
Note: GW = gigawatt-hour.

The differences in regional electricity production trends are largely explained by differences in gross domestic product (GDP) growth rates. As seen in table 2.1, the region with the fastest economic growth between 1971 and 2004 was East Asia and Pacific, where GDP grew by an average of 6 percent per year. The second-fastest growth in GDP occurred in South Asia, where the average annual economic growth rate was 4.7 percent. The difference in GDP growth rates between the Latin America and the Caribbean region and the East Asia and Pacific region was not large in the 1970s. However, the debt crisis in Latin America and the Caribbean during the 1980s led to what is now known as the "lost decade"; growth in East Asia and Pacific has continued at a high and sustained rate.

The electricity generation mix across the world's regions has varied over time. The different electricity production fuel sources considered for this analysis, based on the World Development Indicators database, are (a) coal, (b) hydropower, (c) natural gas, (d) nuclear power, and (e) oil.

Natural gas has become increasingly important for electricity production globally (and in Latin America and the Caribbean) over the past 20 years, gaining ground at the expense of oil and hydroelectric sources. As can be seen in figure 2.2, the share of electricity produced from oil in

Table 2.1   Electricity Production and GDP, Average Annual Growth
percent

	1971–80	1981–90	1991–2000	2001–04
Electricity production				
World	4.8	5.6	2.5	4.2
Latin America and the Caribbean	8.7	5.4	4.3	4.0
East Asia and Pacific	8.3	8.2	8.0	12.7
Middle East and North Africa	12.2	8.1	6.5	7.1
South Asia	7.2	9.3	6.3	5.0
Sub-Saharan Africa	7.8	4.0	2.5	4.4
North America	4.0	2.9	1.7	2.6
GDP				
World	3.8	3.1	2.8	3.1
Latin America and the Caribbean	5.7	1.2	3.3	3.4
East Asia and Pacific	6.6	7.6	8.4	8.9
Middle East and North Africa	5.1	2.8	3.9	4.5
South Asia	3.0	5.4	5.2	7.2
Sub-Saharan Africa	3.7	1.9	2.3	4.9
North America	3.7	3.1	3.1	2.5

Source: Authors' elaboration based on World Development Indicators database 2009.

the world's generation mix declined from 11 percent in 1985 to only 6 percent in 2005. Hydroelectricity experienced a similar decline worldwide, falling from 21 percent to 16 percent. By comparison, the share of electricity produced from hydroelectric sources in Latin America and the Caribbean remains above 58 percent.

Coal has remained the world's primary fuel for electricity generation, and its share has not changed significantly since 1985. A similar trend can be observed in the case of electricity production from nuclear sources: Its share ranged from 16 to 18 percent of total electricity production between 1985 and 2005. Coal plays a dominant role in the generation matrixes of a number of regions (East Asia and Pacific, North America, South Asia, and Sub-Saharan Africa), where electricity production from coal has represented more than 50 percent of total production. When one considers that a number of industrial countries from East Asia and Pacific and from Europe and Central Asia are excluded from the data, coal is also an important source of electricity generation in other countries and regions. For example, in Australia and Japan—which represented 73 percent of electricity production from high-income countries in East Asia and Pacific in 2005—coal accounted for 80 percent

Figure 2.2   World Generation Mix

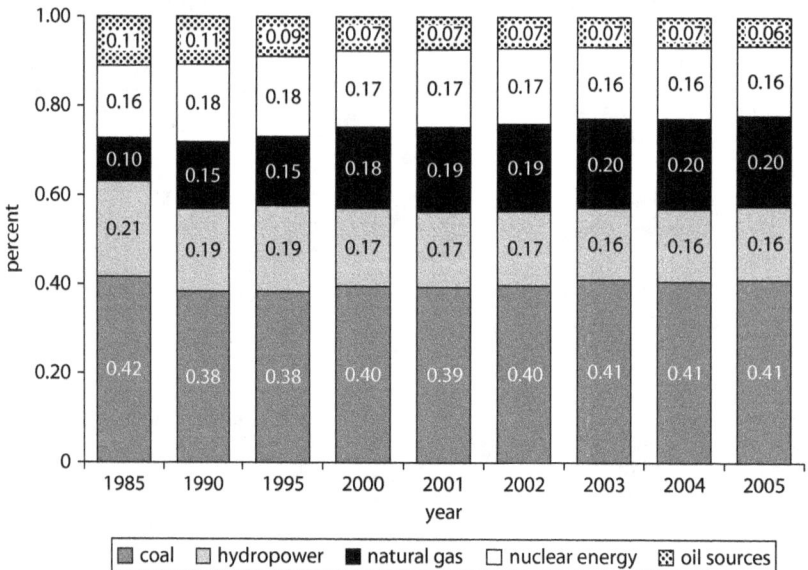

Source: Authors' elaboration based on World Development Indicators database 2009.

and 23 percent, respectively, of power generation.[1] Compared with the global average (41 percent in 2005), the share of electricity production from coal is quite small in Latin America and the Caribbean and has never exceeded 5.6 percent.

Energy Access in Latin America and the Caribbean

The Latin American and the Caribbean region has relatively high electricity access rates compared with other parts of the world. In comparison with Africa's average electricity access rate of 40 percent and the world average of 78 percent, the electricity access rate of Latin America and the Caribbean reached 93 percent in 2008 (table 2.2).

Table 2.2   Electricity Access Rates in Latin America and the Caribbean, 2008

Country	Total (%)	Urban (%)	Rural (%)	Without electricity (millions)
Argentina	97	100	70	1.1
Bolivia	78	98	38	2.2
Brazil	98	100	88	4.3
Chile	99	99	95	0.3
Colombia	94	100	76	3.0
Costa Rica	99	100	98	0.0
Dominican Republic	96	98	90	0.4
Ecuador	92	100	78	1.1
El Salvador	86	97	70	0.9
Guatemala	81	94	68	2.7
Haiti	39	69	12	6.0
Honduras	70	98	45	2.1
Jamaica	92	100	83	0.2
Nicaragua	72	95	42	1.6
Panama	88	94	72	0.4
Paraguay	96	99	88	0.3
Peru	77	96	28	6.5
Trinidad and Tobago	99	100	99	0.0
Uruguay	99	100	86	0.0
Venezuela, RB	99	100	85	0.3
Latin America	93	99	70	34.1
Africa	40	67	23	589.0
World	78	93	63	1,456.0

Source: World Energy Outlook Database 2009.
Note: There are differences in electricity access rates among various database sources. The World Energy Outlook Database is used to ensure consistency in the methodology and to compare countries in Latin America and the Caribbean region with the rest of the world.

Although electricity access rates for the region as a whole are high, there are large inequalities both within and between countries. For instance, nearly 98 percent of Brazil's population has access to electricity, but the access rate in Haiti is only 38 percent.[2] Within countries with high overall access rates, the disparity between urban and rural access rates is startling in some cases. For example, in Peru, which has a national electrification rate of 77 percent, the figure for urban areas is 96 percent, but the electricity access rate in rural areas is only 28 percent. Large disparities between rural and urban electrification rates are also seen in Argentina, Bolivia, Haiti, Honduras, and Nicaragua.

Electricity Capacity and Production Trends in Latin America and the Caribbean

The installed electricity generation capacity in the region has increased from 93 gigawatts (GW) in 1980 to approximately 295 GW in 2008. According to the Latin American Energy Organization (Organización Latinoamericana de Energía, or OLADE) (2009), 53 percent of total electricity generation capacity was hydroelectric, while 44 percent was thermal (coal, natural gas, and petroleum) (see also table 2.3). Nuclear and other types of generating plants accounted for only about 3 percent. By country, the largest producers of electricity in Latin America and the Caribbean were Brazil and Mexico, accounting for almost 56 percent of the total amount of electricity generated in 2008.[3]

The analysis of electricity production trends within Latin America and the Caribbean distinguishes between the following subregions: (1) Brazil; (2) Mexico; (3) Central America: Costa Rica, El Salvador, Guatemala, Honduras, Nicaragua, and Panama; (4) the Caribbean: Barbados, the Dominican Republic, Grenada, Guyana, Haiti, Jamaica, and Trinidad and Tobago; (5) the Southern Cone: Argentina, Chile, Paraguay, and Uruguay;

Table 2.3   Electricity Production in Latin America and the Caribbean
GW

Source	1985	1990	1995	2000	2001	2002	2003
Coal	14,628	22,968	31,947	45,405	46,263	48,473	56,340
Hydropower	311,146	386,434	491,278	584,508	544,483	566,472	585,252
Natural gas	41,590	55,663	77,277	127,866	146,356	163,800	185,747
Nuclear power	9,147	12,455	18,028	20,444	30,064	29,404	31,426
Oil	94,234	114,549	130,614	168,022	167,818	153,250	142,302
Total	470,745	592,069	749,144	946,245	934,984	961,399	1,001,067

Source: Authors' elaboration based on World Development Indicators database 2009.

and (6) the Andean Zone: Bolivia, Colombia, Ecuador, Peru, and República Bolivariana de Venezuela. Within these subregions, there are seven countries—Argentina, Brazil, Chile, Colombia, Mexico, Paraguay, and República Bolivariana de Venezuela—that together accounted for about 89 percent of the region's overall electricity production (figure 2.3).

With more than one-third of the share of total production, Brazil can greatly affect regional generation mix statistics. Specifically, Brazil's reliance on hydroelectricity directly contributes to the high proportion of hydroelectricity in Latin America and the Caribbean, making it the region with the highest share of renewable energy in the world (figure 2.4). Conversely, as the share of electricity contributed by hydroelectricity has fallen in Brazil, the carbon-intensity of electricity production in Latin America and the Caribbean has increased. Following Brazil, Mexico is the second-largest electricity producer in the region, with almost 21 percent of total production in 2005. If Bolivia, Colombia, Ecuador, Peru, and República Bolivariana de Venezuela are treated as a regional cluster—the Andean Zone—their combined electricity production closely follows the

Figure 2.3   Share of Total Electricity Production in Latin America and the Caribbean, 2005

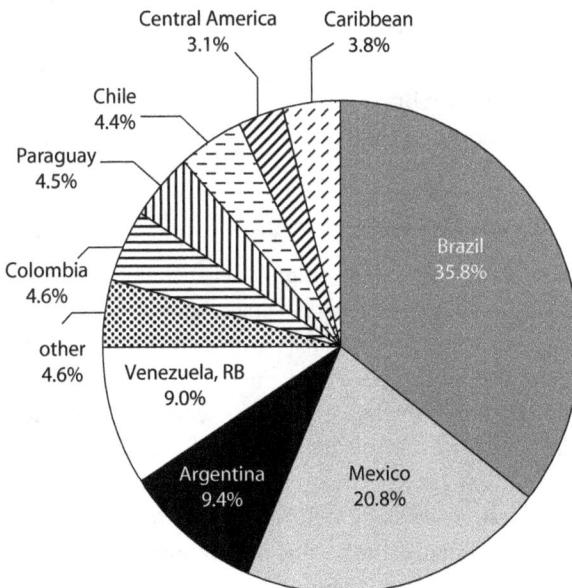

Central America 3.1%
Caribbean 3.8%
Chile 4.4%
Paraguay 4.5%
Colombia 4.6%
other 4.6%
Venezuela, RB 9.0%
Argentina 9.4%
Mexico 20.8%
Brazil 35.8%

Source: Authors' elaboration based on World Development Indicators database 2009.

Figure 2.4   Electricity Production by Subregion, 1985–2005

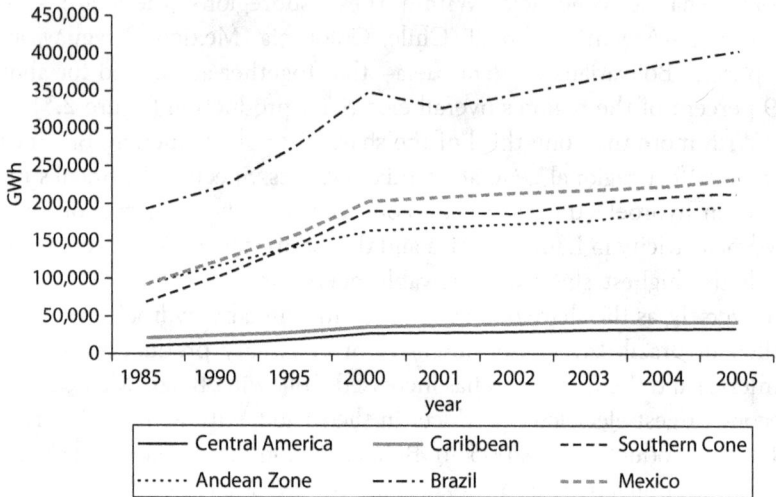

pattern displayed by Mexico and the Southern Cone. The visible change in slope that occurs after 2000 results from the change in time scale used in figure 2.4.[4] There was a slight decline in the region's electricity production between 2000 and 2001, primarily resulting from a drought and other supply problems in Brazil, where hydroelectric production decreased from 349 terawatt-hours (TWh) to 328 TWh.[5]

Regional Generation Mix

The generation mix—that is, the share of electricity production from different power technologies and fuels—has evolved in Latin America and the Caribbean over the past 25 years (figure 2.5). The major changes in the overall generation mix have been a decline in the contribution of oil (diesel and fuel oil), a decline in the share of hydroelectricity (but with hydroelectricity remaining the most important source of electricity), and an increase in the share of natural gas. The share of electricity production from nuclear power and coal has historically been very low in Latin America and the Caribbean and has remained relatively steady since 1985.

The increasing relevance of natural gas as a source for electricity production in Latin America and the Caribbean occurred simultaneously

Figure 2.5   Generation Mix, Latin America and the Caribbean

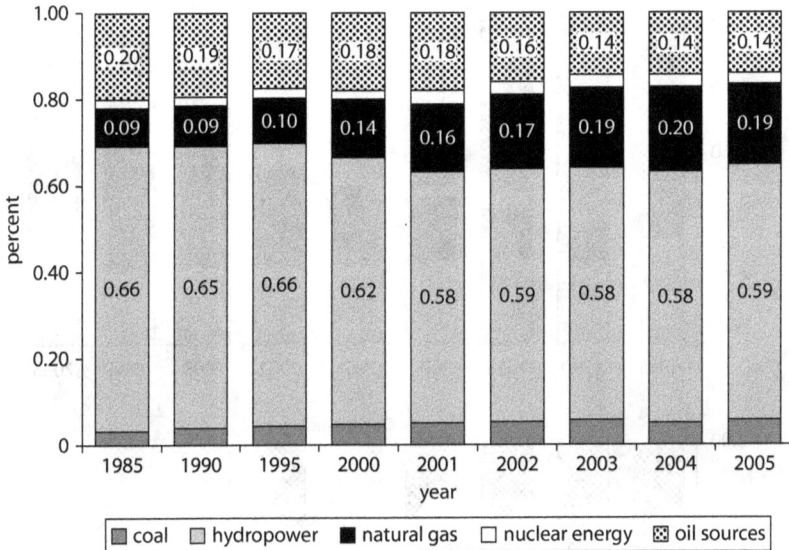

Source: Authors' elaboration based on World Development Indicators database 2009.

with an overall shift away from hydroelectric and oil sources (diesel and fuel oil), which decreased their shares of electricity production from 66 percent to 59 percent and 20 percent to 14 percent, respectively. As shown in figure 2.5, the share of natural gas increased from 9 percent in 1985 to 14 percent in 2000, reaching 19 percent in 2005.

The decline in the role of oil for power generation in Latin America and the Caribbean is reflected quite strikingly in the case of Mexico. As shown in figure 2.6, panel a, in 1985 electricity production from oil sources accounted for 58 percent of Mexico's total production. By 2000, with the rapid increase in new, combined-cycle natural gas plants and the concerted effort by the national utility CFE (Comisión Federal de Electricidad) to close fuel oil–based generating plants, the share of oil had declined to 49 percent, and by 2005, oil accounted for only 31 percent of electricity production in Mexico.

Mexico's relative share of coal-based generation has been compara- tively more volatile in the past few years. Total electricity production between 2002 and 2005 increased steadily, reflecting infrastructure investments made toward the end of the six-year federal political cycle. Of total production, the share of coal-fired electricity experienced major shifts during these four years. Similarly, the trend in the share of natural

Figure 2.6   Generation Mix by Subregion

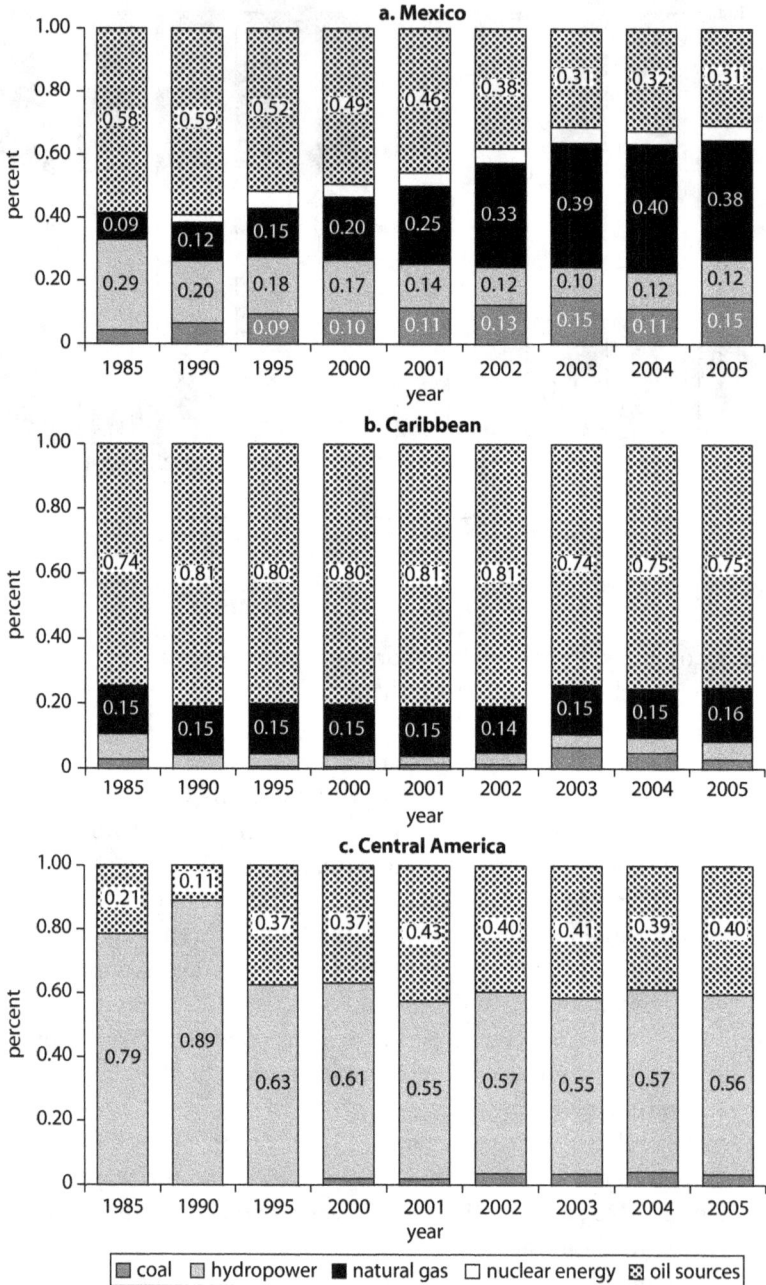

a. Mexico

b. Caribbean

c. Central America

coal   hydropower   natural gas   nuclear energy   oil sources

(continued next page)

Figure 2.6   *(continued)*

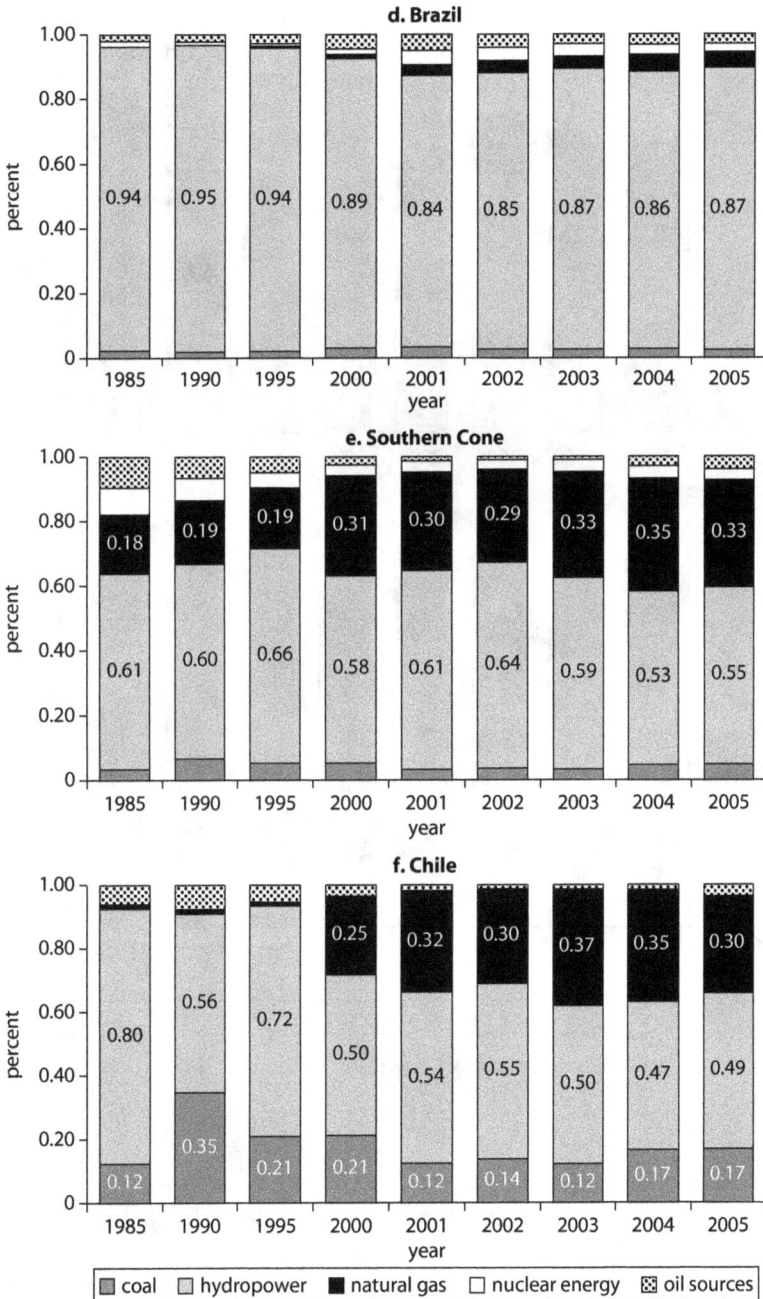

d. Brazil

(percent)

1985	1990	1995	2000	2001	2002	2003	2004	2005
0.94	0.95	0.94	0.89	0.84	0.85	0.87	0.86	0.87

e. Southern Cone

(percent)

1985	1990	1995	2000	2001	2002	2003	2004	2005
0.18	0.19	0.19	0.31	0.30	0.29	0.33	0.35	0.33
0.61	0.60	0.66	0.58	0.61	0.64	0.59	0.53	0.55

f. Chile

(percent)

1985	1990	1995	2000	2001	2002	2003	2004	2005
				0.25	0.32	0.30	0.37	0.35	0.30
0.80	0.56	0.72	0.50	0.54	0.55	0.50	0.47	0.49
	0.35							
0.12		0.21	0.21	0.12	0.14	0.12	0.17	0.17

coal ▧   hydropower ▢   natural gas ■   nuclear energy ▢   oil sources ▨

(continued next page)

Figure 2.6   *(continued)*

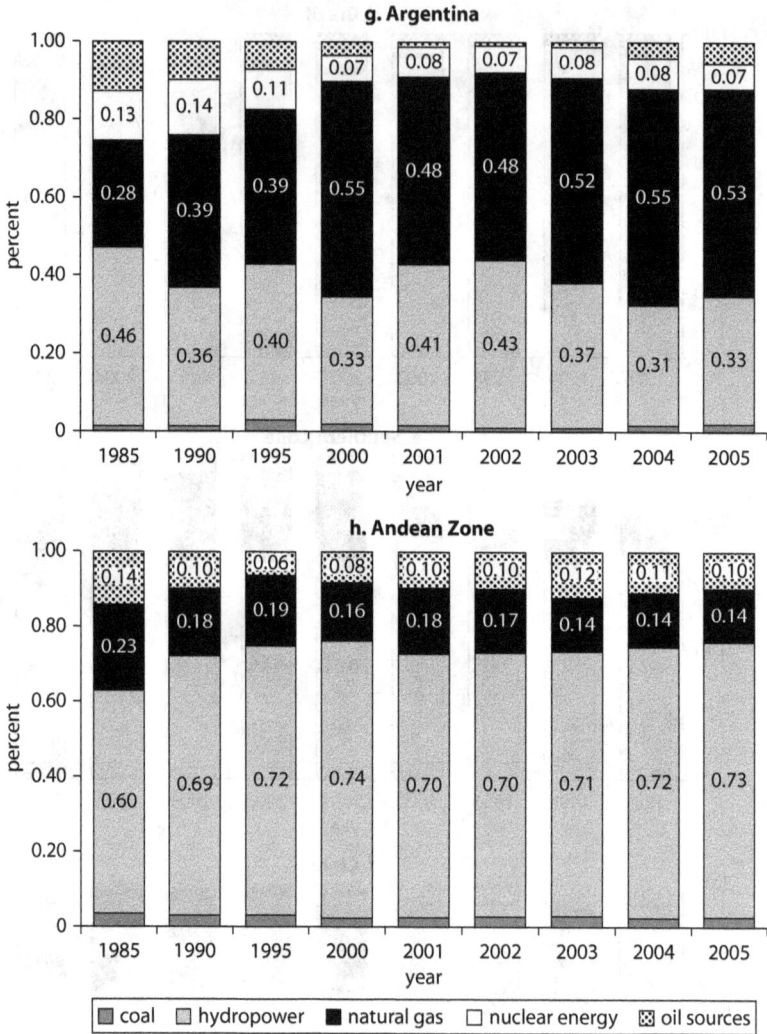

g. Argentina

h. Andean Zone

Source: Authors' elaboration based on World Development Indicators database 2009.

gas–based production was uneven, first increasing abruptly (by 14 TWh) from 2002 to 2003, then slowing to a lower pace (5 TWh) from 2003 to 2004, and, finally, decreasing by 2.3 TWh in 2005.

In the Caribbean subregion, where oil sources play a key role in the overall generation mix,[6] there has been relative stability in the use of oil vis-à-vis other fuels for power generation. As shown in figure 2.6, panel b, the share of oil sources in the electricity production fuel mix has ranged between 74 and 81 percent since 1985. In Central America, electricity has historically been produced mainly from oil and hydroelectric sources (figure 2.6, panel c). However, since 2000, when coal was introduced into the subregion's generation mix, production from oil and hydroelectric sources has experienced a slight decline in relative terms.

From 1985 to 1990, the share of hydropower generation in Central America increased from 79 percent to 89 percent. Between 1990 and 1995, the share of hydroelectricity dropped from 89 percent to 63 percent, with a corresponding increase in power generated from oil products. The drastic changes observed in Central America's energy matrix from 1985 to 1990 and from 1990 to 1995 were mainly driven by changes in Guatemala. Production from hydroelectric sources in Guatemala increased more than threefold between 1985 and 1990, but dropped by 5 percent by 1995.

The increasing influence of natural gas as a source of electricity production in Latin America and the Caribbean is largely the result of the evolution of the generation mix in the Southern Cone, Brazil, and Mexico. The Southern Cone has followed a similar general pattern to that of Mexico. Although starting with a higher share of natural gas in its generation mix relative to Mexico (18 percent compared with Mexico's 9 percent in 1985), by 2005 the Southern Cone had converged with Mexico in terms of the relative importance of natural gas–based generation—33 percent (figure 2.6, panel e). This increase in the overall use of natural gas in the Southern Cone can be explained by developments in two countries—Argentina and Chile, with Argentina showing an annual increase of 1.14 percent during 2000–05 and Chile showing a 1.48 percent annual increase. Although natural gas has become an increasingly important source of electricity generation in both countries, the role of hydro- and coal-based production has declined.

Chile's generation mix, in particular, has changed considerably since 1999. There has been a rapid expansion of natural gas relative to other sources and a steep decline in the share of coal-based production, particularly from 2000 to 2001, when it dropped by nearly 10 percent (figure 2.6, panel f). Unlike in Brazil, the declining role of hydroelectric production in Chile was not caused by climatic conditions. Instead, it can be attributed to the considerable increase in investment in natural gas–based generation facilities, compounded by the lack of investment in hydropower.

In the case of Brazil, although the generation mix has remained comparatively homogeneous with the bulk of the country's electricity generated at hydroelectric plants, natural gas has become more significant since
2000 (see figure 2.6, panel d). In particular, the most notable transition
in Brazil's generation mix occurred from 2000 to 2001, when production
from hydroelectric sources dropped by 5 percent in absolute terms as the
result of a severe drought, while production from natural gas (and nuclear
power) more than doubled.

The absolute levels of production from coal sources during these years
remained almost unchanged. The production of electricity from nuclear
power has been comparatively insignificant in the region's generation mix
through the years, accounting for not more than 3.2 percent since 1985.
The only countries with nuclear plants have been Argentina, Brazil, and
Mexico. In Argentina, nuclear power, although far behind hydroelectric
power and natural gas in its relative importance, has nevertheless been an
important generation source, representing 7 to 8 percent of the total
generation mix since 2000.

Regional Electricity Trade

Cross-border electricity trade and integration is a cross-cutting issue and
has been particularly important for some countries and subregions. A
total of about 56 TWh was traded (that is, the aggregate of power
exchanges between countries) by the region in 2006, of which 41.5 TWh
were accounted for by Brazil. Of the region's total electricity exports of
about 54 TWh in 2006, Paraguay was the single-largest trading country,
accounting for 46 TWh (primarily exports from the large Itaipú and
Yacyretá hydropower plants). Electricity trade occurs primarily in three
separate zones: (a) Argentina, Brazil, Paraguay, and Uruguay; (b)
Colombia, Ecuador, Peru, and República Bolivariana de Venezuela; and
(c) Mexico and Central America.

Table 2.4 presents historical data on electricity imports and exports in
Latin America and the Caribbean. Aggregating the data for all countries
in the region, table 2.4 shows that the resulting trade balance[7] does not
equal zero, which is explained by Mexico's electricity trade "outside" the
region with the United States. The data show a slowdown in trade after
2000, reflecting a growing trend of countries to supply domestic needs
first.[8] Paraguay plays the most important role in the region's electricity
trade, being the largest net exporter. With an average of 45 TWh exported
annually since 2000, Paraguay is a major electricity supplier to Brazil and

Table 2.4 Electricity Imports and Exports

GWh

	Product	1985	1990	1995	2000	2001	2002	2003	2004	2005	2006	2007
Argentina	Imports	—	32	220	6,023	5,662	2,856	2,543	4,144	4,140	6,193	2,628
	Exports	2,674	2,682	2,343	7,250	7,417	8,776	7,579	7,613	8,018	7,418	10,275
Brazil	Imports	5	7	—	7	6	7	6	7	160	283	2,034
	Exports	1,918	26,538	35,343	44,333	37,844	36,570	37,141	37,392	39,202	41,447	40,866
Chile	Imports	—	—	—	—	—	—	—	—	—	—	—
	Exports	—	—	—	—	—	—	—	—	—	—	—
Colombia	Imports	—	—	—	1,190	1,386	1,813	1,667	1,774	2,152	2,285	1,628
	Exports	—	—	—	37	210	618	1,182	1,682	1,758	1,813	877
Ecuador	Imports	—	200	370	77	40	8	69	48	16	21	39
	Exports	—	—	—	—	—	—	67	35	16	1	39
Mexico	Imports	237	1,946	1,944	195	22	56	1,120	1,642	1,723	1,570	861
	Exports	140	575	1,164	1,069	271	344	954	1,006	1,291	1,299	1,451
Paraguay	Imports	40	48	—	—	327	531	71	47	87	523	277
	Exports	2,860	24,797	35,369	47,331	39,109	41,770	45,173	45,003	43,784	45,706	45,133
Peru	Imports	—	—	—	—	—	—	n.a.	—	2	1	—
	Exports	—	—	—	—	—	—	—	—	8	—	—
Venezuela, RB	Imports	2,678	2,589	233	942	1,377	2,287	1,138	1,138	841	16	995
	Exports	—	51	188	1,328	123	559	434	2,348	1,585	2,835	788
Central America	Imports	212	512	354	1,493	877	965	851	1,257	560	261	305
	Exports	229	497	372	1,484	944	969	845	1,207	565	275	307

Sources: OLADE and Sistema de Información Económica Energética (SIEE) (demand and supply module).

Note: GWh = gigawatt hour, — = not available.

Argentina. Brazil's demand for electricity from international trade can be inferred from table 2.4, which shows the difference between the countries' own production and domestic consumption. In 2007, this difference was about 39 TWh, a figure equivalent to Brazil's trade balance (exports minus imports) for that year. Brazil has structured its electricity supply system taking into account imports from Paraguay.

Trade between Mexico and the countries in Central America is expected to grow with the completion of the SIEPAC (Sistema de Interconexión Eléctrica para América Central, or Central American Electrical Interconnection System) network. To date, Mexico has been exporting an average of about 1 TWh annually over the past 20 years; at the same time, Mexico's imports relative to domestic production have remained insignificant.

In Central America, power interconnections currently exist between each of the countries, in both directions. In terms of capacity, the largest interconnection is from Panama to Costa Rica (110 megawatts [MW]), while the smallest interconnection is from Costa Rica to Nicaragua (60 MW). Table 2.5 presents the complete list of interconnections within Central America.

Map 2.1 illustrates the interconnections currently in operation, as well as those under construction or being planned or studied. Argentina, Brazil, Paraguay, and Uruguay have the most interconnections currently in operation. The figure also indicates the planned interconnection lines between Brazil and Uruguay. New interconnections are been studied between Colombia and Panama, and Bolivia and Peru.

Colombia stands out as an important net exporter, with export volumes becoming increasingly significant since 2000. Ecuador has been

Table 2.5   Electricity Interconnections in Central America, 2006

Export country	Import country	Capacity (MW)
Guatemala	El Salvador	100
El Salvador	Guatemala	95
El Salvador	Honduras	100
Honduras	El Salvador	100
Honduras	Nicaragua	80
Nicaragua	Honduras	80
Nicaragua	Costa Rica	60
Costa Rica	Nicaragua	60
Costa Rica	Panama	70
Panama	Costa Rica	110

Source: Comisión Regional de Interconexión Eléctrica 2007.

Map 2.1   South American Interconnections

● ━━ ● interconnection in construction   ● ┄┄ ● interconnections currently in operation

◌ ┄┄ ◌ planned interconnection   ◌ ┄┄ ◌ studies being conducted

Source: Authors' based on CIER 2008; potential interconnections included by Alberto Brugman.

the main recipient of Colombian electricity exports. In addition to its trade with Ecuador, Colombia has been engaged in electricity trade with República Bolivariana de Venezuela, which, at the same time, has also provided electricity to Brazil's north. Figure 2.7 presents the electricity production and electricity demand across the countries of the region in 2007.

Figure 2.7   Electricity Production and Domestic Demand in Latin America and the Caribbean, 2007

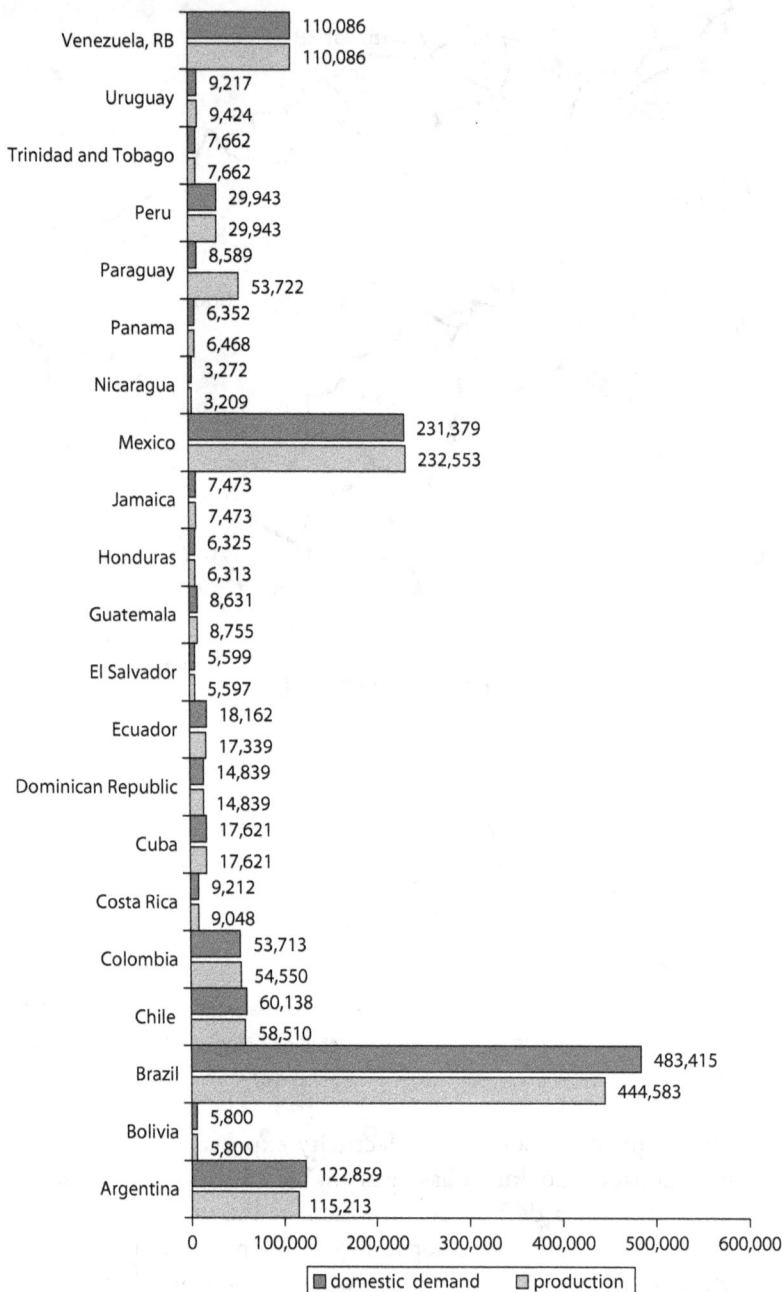

Source: Authors' elaboration based on OLADE, SIEE (demand and supply module).
Note: GWh = gigawatt hour.

Power Sector Structure in Latin America and the Caribbean

The electricity sectors in different Latin American countries have distinct regulatory and market structures. Some countries have a completely vertically integrated and state-owned structure, such as Costa Rica, Ecuador, and Paraguay. At the opposite end of the spectrum, the power sectors in Chile and Panama have fully market-oriented structures with private sector participation in all segments of the market. Other countries currently fall somewhere in between, although the structure of the power sector in many countries has been in a constant state of flux over the past 30 years. During the economic crisis of the 1980s, private sector investment retreated from the power sector in many Latin American and Caribbean countries. To guarantee the provision of service, many governments nationalized the power sector, giving state-owned utilities vast control over electricity markets. During the 1980s and 1990s, the growing electricity capacity requirements and the lack of private sector investment led to a strain on public utilities. This in turn led some governments to provide incentives and a competitive investment climate for private sector companies in certain segments of the market, including generation in some countries and distribution in others.

Economic liberalization of the power sector in Latin America began in Chile in 1982 with the privatization of its utility companies, the creation of a spot market, and the opening of the sector to new investors. Following the Chilean experience, many Latin American countries introduced a range of similar policies to restructure their electric power sectors. New, independent regulatory agencies were created; large state-owned companies were unbundled and privatized; and competitive market-oriented frameworks were implemented throughout the 1990s in a range of countries. The process continues today.

In Mexico, liberalization in the 1990s opened the generation segment of the market to large independent power producers, mainly building and operating combined-cycle natural gas plants. In Central America, reforms were implemented in El Salvador, Guatemala, Nicaragua, and Panama, which liberalized their entire electricity markets. In Costa Rica and Honduras, the reforms were limited to the opening of the generation segment. In South America, the most extensive reforms—new liberalized markets with significant private participation—were introduced in Argentina, Bolivia, Brazil, Colombia, and Peru. Ecuador introduced a competitive wholesale electricity market, although it has not been opened to significant private participation. The electricity markets of Paraguay and República Bolivariana de Venezuela have remained largely unchanged with a dominant public sector presence. Chile kept in place its already privatized market.

New regulatory frameworks have redefined the conditions for electricity service in most Latin American countries, frequently providing for a structure under which the role of the state is limited to the formulation of policies, the exercise of regulatory functions for their respective power sectors, and the administration of concessions. However, in most countries the state remains an important player in the sector through its ownership of companies involved in generation and in transmission and generation (including, to varying degrees, countries such as Brazil, Colombia, the Dominican Republic, Guatemala, Mexico, and Uruguay).

Impact of the Financial Crisis in GDP Forecast

Since the last quarter of 2008, forecasts for the majority of macroeconomic variables have changed as a result of the international financial crisis. Changes in economic growth forecasts have major implications for the growth in electricity demand, and the forecasts of GDP have been modified frequently over the course of preparing this report.

The June 2008 average GDP growth forecast for Latin America and the Caribbean was 3.9 percent, indicating that no major changes were foreseen in the summer before the start of the financial crisis. In August 2008, the GDP forecasts for most countries of the region were positive and followed the trend of stable economic growth that had characterized Latin America and the Caribbean during the previous five years. The February 2009 GDP growth forecast was 3.7 percent, according to the Latin American consensus.[9]

Given the long-term outlook of this report, experience shows that in the context of a time horizon greater than 20 years, any slowdown in demand and supply of electricity in the short term is likely to be made up over time. The International Monetary Fund (IMF) GDP growth assumptions used in this report are taken from the *World Economic Outlook, April 2009* (IMF 2009). According to those projections, global GDP was forecast to decrease by 1.3 percent in 2009, but then recover with an overall increase of 1.9 percent in 2010. The GDP of emerging and developing economies as a whole is forecast to grow at 4.0 percent in 2010, but among Latin American and Caribbean countries, the projected GDP growth in 2010 is only 1.6 percent. The two largest countries in the region, Mexico and Brazil, are forecast to have GDP growth in 2010 of 1.0 percent and 2.2 percent, respectively, with the longer-term projections (through 2014) looking more optimistic with 4.9 percent and 4.5 percent, respectively. It should be noted that a mid-year update of key World Economic Outlook projections was released by the IMF in July 2009 and included modest adjustments to the GDP growth forecasts. This report relies on the IMF's April 2009 GDP growth projections, because the July 2009 update does not provide country-by-country projections and because over the longer term (through 2030), the short-term bumps in the next two years will not have a major effect.

Notes

1. Furthermore, Germany and the United Kingdom use 50 percent and 35 percent, respectively, of coal in their generation mixes. Similarly, half of the electricity production in the United States is coal based.

2. Brazil has expanded access to electricity in rural areas through a program known as Luz Para Todos. Begun in 2003, the program is a partnership between the federal government, state agencies, and distribution companies.

3. Electricity generation capacity refers to the total installed capacity capable of generating electricity. It differs from electricity generation, which refers to the actual electricity production in a given period of time. There is often a significant difference between a country's installed capacity and power generation from a specific source, depending on resource availability (such as water flow), maintenance, and other variables affecting the "plant factor."

4. In figure 2.4, the scale is for 5-year periods from 1985 until 2000. After 2000, the scale is annual.

5. Between June 2001 and February 2002, Brazil was forced to ration electricity, but there were no forced outages. The rationing program was quite successful in reducing demand, achieving nearly a 20 percent reduction in consumption in a very short time. The experience demonstrates that consumers can react to price and other signals. Over the longer term, permanent improvements in efficiency were only about one-quarter of the temporary reduction in consumption that was achieved.

6. Although power generation in Caribbean countries is dominated by oil, countries in the subregion with the highest percentage contribution from hydroelectric sources are the Dominican Republic and Haiti.

7. According to OLADE's figures, domestic supply equals the country's total production, minus exports, plus imports.

8. Trade reemerged in 2004.

9. The survey date was June 16, 2008, with the disclaimer made by the Latin American consensus that its estimates were based on surveys of 120 prominent Latin American and Caribbean economic and financial forecasters.

References

CIER (Comisión de Integración Energética Regional). 2008. "Síntesis Informativa Energética de los Países de la CIER 2007." CIER, Montevideo, Uruguay.

IMF (International Monetary Fund). 2009. *World Economic Outlook, April 2009: Crisis and Recovery*. Washington, DC: IMF.

OLADE (Organización Latinoamericana de Energía). 2009. *Informe de Estadísticas Energéticas 2009, Año Base 2008*. Quito: OLADE.

CHAPTER 3

Baseline Electricity Supply Scenario to 2030

As seen in the previous chapter, the Latin America and the Caribbean region has been quite successful in expanding the supply of electricity and in achieving relatively high electricity access rates. Because of the predominance of hydroelectricity, the region as a whole has the lowest carbon-intensity of any region in the world, whereas the countries with large amounts of hydroelectricity have benefitted from relatively low and stable electricity prices compared with those relying on fossil fuels. Going forward, the region will need to respond to the electricity challenges set out in chapter 1, specifically: (a) meeting future demand growth, (b) maintaining the security of energy supply, (c) minimizing cost and maximizing efficiency, (d) limiting environmental impacts, and (e) putting in place the necessary policy and regulatory regime to achieve (a) through (d).

The main purpose of this chapter is to present electricity demand and supply scenarios for the Latin America and the Caribbean region to 2030. These scenarios reflect the power expansion plans currently available in countries, and can therefore be viewed as the baseline for the region going forward. The supply-side scenarios illustrate the collective future of the generation mix for the region. The first part of this chapter describes the specific assumptions and tools used in the construction of supply scenarios. The second part of the chapter presents the results of the scenario analysis for individual countries and subregions. The analysis identifies cost-minimizing investment and production strategies to meet the projected demand as well as the amount of investment required to implement such a strategy. The third part of the chapter presents estimates of carbon dioxide (CO_2) emissions followed by a brief sensitivity analysis that examines the effect of carbon taxes on the generation mix.

Modeling Framework

For illustration of the implications of current trends in electricity development—for individual countries, subregions, and the region as a whole—scenarios of electricity demand and supply to 2030 were created with a simple electricity demand function and a detailed energy supply planning model. The resulting scenario, referred to as the ICEPAC (Illustrative Country Expansion Plans Adjusted and Constrained) scenario, reflects the current expansion plans of individual countries in the region, to the extent that information was available.

Demand Function

To estimate electricity demand, we developed a log-linear model using gross domestic product (GDP) and electricity prices as explanatory variables. The demand model makes use of energy statistics from the database of the Latin American Energy Organization (Organización Latinoamericana de Energía, or OLADE). This database is the largest and most complete set of energy data for the region. With the help of OLADE, an electricity demand scenario for the region was created to 2030.

The estimation of electricity demand uses the most simplified specification for electricity demand considering GDP and electricity prices.[1] The demand scenario for each country was generated using the GDP demand elasticity and a forecast for GDP up to 2030. For simplification of the demand scenario, a constant GDP growth rate of 3 percent between 2015 and 2030 was assumed. An additional simplifying assumption was that real electricity prices are constant over the period.[2] Given the long-term nature of the exercise, the preference was to follow the historical trends in electricity demand, rather than to forecast the expected changes in one or more explanatory variables. Historical data from 1978–2007[3] were used to approximate a log-linear regression, where α, β, and γ are the parameters to be estimated. In the following regression, β represents the long-term GDP-electricity demand elasticity, and γ represents the long-term electricity price elasticity of demand.[4]

Ln (total electricity demand) = α + β Ln (gross domestic product) + γ Ln (electricity price)

GDP Growth Rate and Electricity Price Assumptions

The GDP data used were taken from the International Monetary Fund's (IMF) World Economic Outlook Database (April 2009) (IMF 2009). The database contains the forecast for each country's GDP growth rate to 2014. Table 3.1 contains the annual GDP growth assumptions by country that were used in the model. For GDP forecasts from 2014 to 2030, a constant annual GDP growth rate of 3 percent was assumed for all countries. As noted, a constant real electricity price was assumed for all countries of the region.[5]

The results derived for the GDP and price coefficients (table 3.2) are fully consistent with economic theory: Demand for electricity is positively correlated with income and negatively correlated with price. The higher the income of a country, the more power the country consumes. As

Table 3.1   GDP and Price Coefficients

Country	GDP	Price
Argentina	1.02	−0.78
Bolivia	2.06	−0.62
Brazil	1.55	−0.91
Chile	1.14	0.03
Colombia	1.17	−0.36
Costa Rica	1.26	−0.84
Ecuador	1.60	−1.06
El Salvador	1.52	−0.55
Guatemala	1.45	−0.70
Honduras	1.66	−1.67
Mexico	1.28	−0.76
Nicaragua	2.24	−0.36
Panama	1.29	−0.36
Paraguay	0.95	−1.38
Peru	1.13	−0.72
Uruguay	1.64	−1.38
Venezuela, RB	0.48	−0.05

Source: Authors' elaboration.
Note: GDP coefficients were statistically significant for all countries. Price coefficients were not statistically significant for Chile, Colombia, Nicaragua, and República Bolivariana de Venezuela. The results derived for the GDP and price coefficients are fully consistent with economic theory: GDP growth is positively correlated with electricity demand, and electricity prices are negatively correlated.

Table 3.2   IMF World Economic Outlook Database GDP Forecasts, Constant Prices
annual % change

Country	2008	2009	2010	2011	2012	2013	2014
Argentina	7	-2	1	3	3	3	3
Barbados	1	-4	1	3	3	3	3
Bolivia	6	2	3	3	3	3	4
Brazil	5	-1	2	3	3	4	4
Chile	3	0	3	4	5	5	5
Colombia	3	0	1	4	5	5	5
Costa Rica	3	1	2	4	5	5	5
Dominican Republic	5	1	2	5	7	7	7
Ecuador	5	-2	1	2	2	3	3
El Salvador	3	0	1	2	4	4	5
Guatemala	4	1	2	4	4	4	4
Haiti	1	1	2	3	3	3	4
Honduras	4	2	2	2	3	3	3
Mexico	1	-4	1	5	5	5	5
Nicaragua	3	1	1	2	3	4	4
Panama	9	3	4	7	7	7	7
Paraguay	6	0	2	3	4	5	5
Peru	10	3	4	7	6	6	6
Suriname	7	3	3	4	5	5	5
Trinidad and Tobago	3	1	2	3	3	3	3
Uruguay	9	1	2	4	4	4	4
Venezuela, RB	5	-2	-1	1	1	1	1

Source: IMF 2009.

countries develop, the share of the industrial and service sectors (which consume relatively large amounts of power) in the economy rises, while the share of agriculture (which uses relatively little electricity) tends to fall. Also, higher household income levels are associated with higher electricity consumption, reflecting the increase in the use of durable goods.

The GDP-electricity demand elasticity is always greater than unity, except in the case of Paraguay and República Bolivariana de Venezuela. Whereas faster GDP growth may be associated with higher electricity demand, the GDP-electricity elasticities are a measure of energy intensity, because they represent the percentage increase in electricity demand for every percentage increase in GDP. The lower the elasticity, the less power is required per percentage point of GDP growth. However, although intensity can be attributed to the efficient use of electricity, it also depends on the structural composition of the economy. For the region as a whole, a one percentage point increase in GDP on average results in a 1.37 percent increase in electricity consumption.

Despite the limited analysis of the effect of electricity prices, estimates confirm that the higher the price, the lower the demand for electricity. Price elasticity estimates show that the demand for electricity is inelastic in most countries, with most elasticity estimates being less than one. Except for Ecuador, Honduras, Paraguay, and Uruguay, the demand for electricity falls by less than 1 percent for every one percentage point increase in price.

For reference, box 3.1 provides an overview of the various ways in which other models have attempted to incorporate the income and price elasticity of demand into electricity demand forecasts. Appendix B also

Box 3.1

Price and Income Elasticity of Electricity Demand

As summarized in the figures below, academic research shows substantial variation in the estimates of the price and income elasticity of electricity demand. The short-run and long-run values are also displayed below:

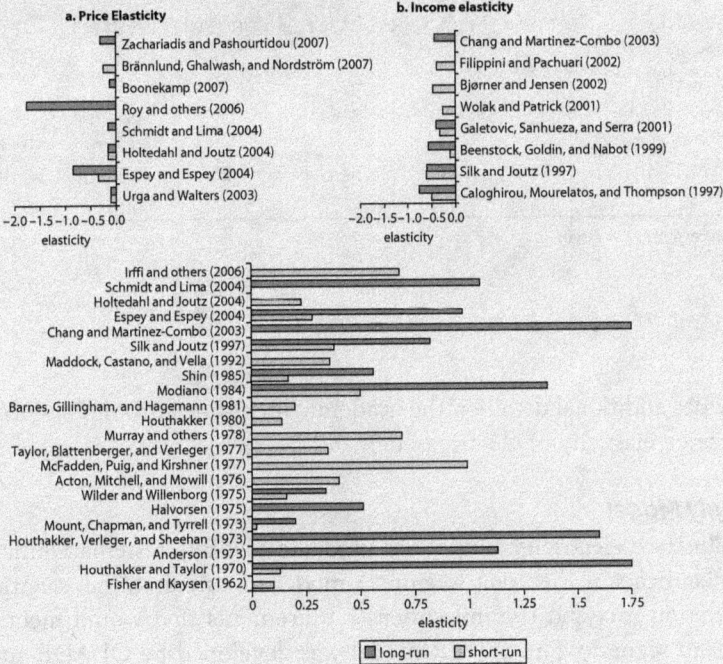

a. Price Elasticity

Zachariadis and Pashourtidou (2007)
Brännlund, Ghalwash, and Nordström (2007)
Boonekamp (2007)
Roy and others (2006)
Schmidt and Lima (2004)
Holtedahl and Joutz (2004)
Espey and Espey (2004)
Urga and Walters (2003)

elasticity

b. Income elasticity

Chang and Martinez-Combo (2003)
Filippini and Pachuari (2002)
Bjørner and Jensen (2002)
Wolak and Patrick (2001)
Galetovic, Sanhueza, and Serra (2001)
Beenstock, Goldin, and Nabot (1999)
Silk and Joutz (1997)
Caloghirou, Mourelatos, and Thompson (1997)

elasticity

Irffi and others (2006)
Schmidt and Lima (2004)
Holtedahl and Joutz (2004)
Espey and Espey (2004)
Chang and Martinez-Combo (2003)
Silk and Joutz (1997)
Maddock, Castano, and Vella (1992)
Shin (1985)
Modiano (1984)
Barnes, Gillingham, and Hagemann (1981)
Houthakker (1980)
Murray and others (1978)
Taylor, Blattenberger, and Verleger (1977)
McFadden, Puig, and Kirshner (1977)
Acton, Mitchell, and Mowill (1976)
Wilder and Willenborg (1975)
Halvorsen (1975)
Mount, Chapman, and Tyrrell (1973)
Houthakker, Verleger, and Sheehan (1973)
Anderson (1973)
Houthakker and Taylor (1970)
Fisher and Kaysen (1962)

elasticity

■ long-run □ short-run

(continued next page)

Box 3.1 (*continued*)

However, in the analysis of the variations in price and income elasticity, it is important to identify the country, time period, and sector in which the study was conducted to ensure a fair comparison. The table below provides additional details on the Latin America and the Caribbean specific studies.

Box table 3.1   Price and Income Elasticity of Electricity Demand

Source	Country	Time period	Sector	Elasticity type	Term of study	Finding
Irffi and others (2006)	Brazil	1970–2003	Residential	Income	Short run	0.84
Schmidt and Lima (2004)	Brazil	1980–2000	Residential and industrial	Income	Short run	1.10
			Industrial	Price	Long run	–0.13
			Residential			–0.15
Chang and Martinez-Combo (2003)	Mexico	1985–2000	Residential	Income	Long run	1.95
				Price		–0.44
			Industrial	Income		1.29
				Price		–0.25
Maddock, Castano, and Vella (1992)	Colombia	1986	Residential	Income	Short run	0.33
				Price		–0.32
Galetovic, San-hueza, and Serra (2001)	Chile	2001	Residential	Price	Short run	–0.33
					Long run	–0.41
			Commercial		Short run	–0.19
					Long run	–0.21

Source: Authors' elaboration.

provides additional detail on the academic literature surrounding income and price elasticity of electricity demand.

Supply Model

To illustrate electricity supply, we used the SUPER (Sistema Unificado de Planificación Eléctrica Regional) model to calculate the electricity generation mix and the investment requirements that would meet the demand scenario. The SUPER model was developed by OLADE and is aimed at prioritizing, scaling, and selecting electricity projects to meet the growth in electricity demand. In each phase, the model determines generation targets for each of the system's power plants; minimizes the expected value of the operating and capital costs throughout the period; and estimates selected environmental impacts, such as CO_2 emissions, associated with future development of the electric power sector. After the total annual demand is calculated, the supply model is used to determine the optimal, cost-minimizing generation mix to meet the demand.

In addition to the demand scenario, various data are also inputted in the SUPER model, including hydrology, fuel reference prices, existing power plants with their operational features, projects under construction and their entry dates, as well as eligible projects with entry dates, investment costs, and operational variables.

Advantages and Constraints of the Supply Model

The SUPER model has several advantages for simulating electricity supply systems. It allows finding minimum cost-generation alternatives that meet the demand requirements in an electricity system. In addition, the SUPER model allows for an evaluation of the generation expansion considering both renewable and thermal technologies. For this study, the SUPER model was used to calculate the electricity generation mix considering the minimum cost of electricity depending on the capital and variable costs of the electricity generation technologies available in the country. The capital and variable costs of technologies used to calculate the supply-side optimal technology mix were fixed across the region. For example, the variable and fixed costs to generate a unit of electricity (gigawatt-hour, or GWh) with a combined-cycle technology were the same for all countries in the region. The generation technologies used in the model considered a country's available capacity in 2007 and the information available from the country's electricity expansion plan. One limitation of this approach is that it does not reflect potential large-scale changes in the base-case country expansion plans (such as potential shifts toward a larger proportion of cleaner technologies as a result of concessional climate funds or carbon taxes).

Although the SUPER model allows for an integral evaluation of the electricity generation in a country, the information available limits the scope of the analysis. For a more precise estimate of the electricity generation mix in the coming years, the model is based on information from the electricity expansion plans of the countries in the region. The more information available on the technologies and fuel sources of a specific country, the more accurate the forecast of future generation composition.

ICEPAC Scenario

Applying the outputs from the demand and supply models, this report presents the ICEPAC scenario. This scenario was created using the results of the demand model, which are then used as an input to OLADE's SUPER model, which is used to calculate the optimal, least-cost generation mix that would satisfy the demand. The ICEPAC scenario is thus "Illustrative" of the current power planning within the region because it is based on (a) "Country Expansion Plans" to 2030 (where available), which are then (b) "Adjusted" to account for the lack of data and to extrapolate country expansion plans (most of which are available to 2018 or 2020), and then (c) "Constrained" so as not to exceed energy resource potential (such as domestic hydroelectric resources) but to allow use of a database of international technology supply costs, which places a cost-minimizing constraint on the electricity supply model. From the ICEPAC scenario, it is possible to compare electricity supply with demand, estimate the financial needs for meeting new supply, and provide a baseline with which to compare alternative means for meeting supply and regulating demand.

The ICEPAC scenario for the expected generation mix takes into account a variety of factors:

- *Fuel prices.* Fuel prices are one of the main factors affecting the composition of the generation mix. Fuel-based technologies have high variable costs and low capital costs compared with renewable technologies; thus, in the long term, a comparison of the fuel costs of different technologies is crucial to estimating the supply mix.
- *Resource availability.* Resource availability also affects the technology composition, because it may impose a constraint on the feasibility of each country's best economic alternative.
- *National expansion plans.* Country-level generation plans are also an important factor to consider when forecasting the future generation mix.

However, an important limitation of the ICEPAC scenario is that it is based on current country expansion plans, namely those in existence before 2009. Although these plans provide a uniform basis for projecting future expansion, many do not reflect the contribution of non-hydro renewables, electricity trade, and energy efficiency.[6] These shortcomings of the modeling analysis are addressed in chapter 4.

Modeling Results

This section presents the results of the demand modeling and the resulting ICEPAC scenario. The first part presents the electricity demand scenario by subregion. The second part presents the ICEPAC supply scenarios, for the region and each subregion, and the corresponding generation mix of technologies through 2030. In consideration of the long time horizon of the study, the interpretations of the results are focused more on the general forecasted trends and patterns in supply, demand, and generation mix, rather than on individual numbers. The remaining sections of this chapter explore (a) the amount of additional investment that will be needed on a regional and a subregional level to meet the ICEPAC scenario generation volume and specific technological mix, and (b) the estimated CO_2 emissions for the region and each subregion through 2030 considering the fuel mix forecasted in the generation matrix.

Electricity Demand

The electricity sector in Latin America and the Caribbean will experience substantial growth over the next 20 years. According to the electricity demand modeling exercise conducted, the area's total demand for electricity will reach nearly 2,500 terawatt-hours (TWh) in 2030, approximately twice the 2008 level (figure 3.1). From 2008 to 2014, the average annual growth is estimated to be approximately 3.7 percent, with total demand increasing nearly 22 percent by 2015 (table 3.3). A subsequent 78 percent increase in demand will occur from 2015 to 2030. The share of each subregion's electricity production is estimated to remain roughly the same throughout the 2008–30 timeframe. This is to be expected, given the average annual GDP growth assumptions used in the calculations.

Although no one particular subregion distinctly dominates the cumulative increase in electricity demand, electricity demand in Central America increases at a slightly faster pace than in other subregions, with an average annual growth in demand of 5.3 percent. Brazil and Mexico, with their annual electricity demand growing at 4.7 percent and

Figure 3.1   Electricity Demand by Subregion

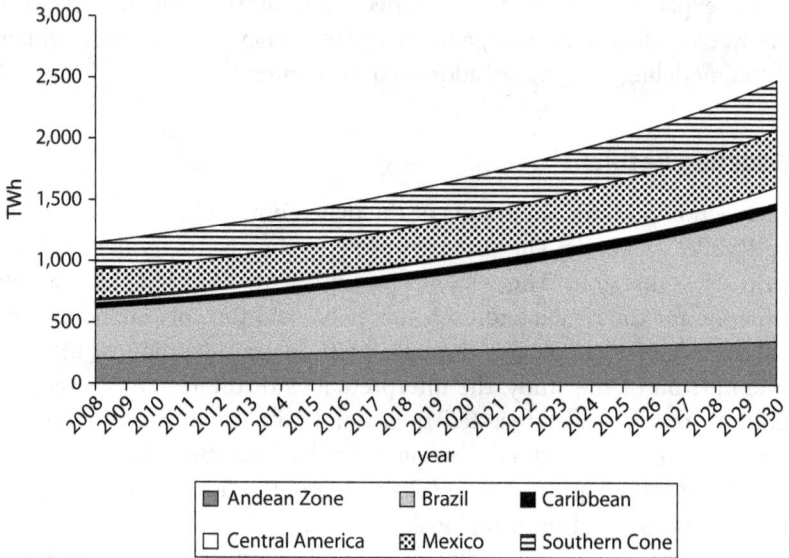

Source: Authors' elaboration based on optimization model.

Table 3.3   Electricity Demand by Subregion, 2008–30

percent

Subregion	Average annual growth
Andean Zone	2.8
Brazil	4.7
Caribbean	3.2
Central America	5.3
Mexico	3.4
Southern Cone	2.8
Average	**3.7**

Source: Authors' elaboration.

3.4 percent, respectively, together represent nearly 62 percent of the region's total demand in 2030, whereas the Andean Zone and the Southern Cone countries comprise another 31 percent.

Supply Side: ICEPAC Scenario

Estimates for the region are presented, followed by those for the subregions. The analysis shows the level and generation mix for the region and the subregions up to 2030. The modeling exercise indicates the supply forecast for different power generation technologies across the region and their share in total generation.

Level and generation mix for Latin America and the Caribbean: In the Latin America and the Caribbean region as a whole, under the ICEPAC modeling exercise it is estimated that an additional 239 gigawatts (GW) of installed capacity will be required to satisfy the estimated demand. To meet this demand, the total electricity generation mix in Latin America and the Caribbean by 2030 under baseline economic conditions is expected to be dominated by hydroelectricity and natural gas, with shares of 50 percent and 30 percent, respectively. The future technology mix is not expected to change significantly from 2008 onward (figure 3.2). The main change estimated by the model is a slight decrease (of about 9 percent) in the share of hydroelectric generation, which will occur as a result of expansion in natural gas and coal, whose respective shares in the total generation mix will increase by 7 percent and 3 percent, respectively (table 3.4).

Table 3.4 shows the regionwide ICEPAC scenario generation mix from 2008 to 2030. The modeling exercise indicates that Latin America and the Caribbean will experience a modest decline in the share of hydropower (from 59 percent to 50 percent) and a steep decline in the share of fuel oil (from 8 percent to 3 percent), compensated by an increase in

Figure 3.2   Regionwide Electricity Generation Mix, 2008–30

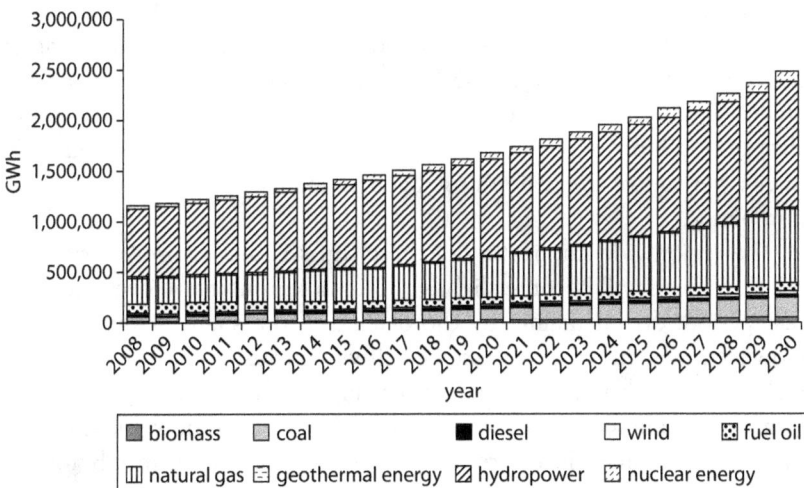

Source: Authors' elaboration based on optimization model.

Table 3.4   Regionwide Electricity Generation Mix, 2008–30
percent

Source	Mix, 2008	Mix, 2030
Biomass	0.5	2.0
Coal	4.6	7.9
Diesel	2.3	1.2
Wind	0.1	1.3
Fuel oil	8.4	3.3
Natural gas	22.0	29.4
Geothermal energy	1.0	0.8
Hydropower	58.6	50.0
Nuclear energy	2.8	4.2

Source: Authors' elaboration based on optimization model.

the share of coal by approximately 3 percent (from 5 percent to 8 percent), an increase in the share of natural gas by 7 percent (from 22 percent to 29 percent), and a moderate increase in nuclear energy (from 3 percent to 4 percent).

Changes in the relative shares of each generation source are a result of a relatively small compounded annual growth in hydroelectric generation (of only 2.8 percent) against the annual growth rates projected for gas and coal technologies of 5.1 percent and 6.1 percent, respectively. Hydroelectric energy will thus continue to be the most important source of generation in the region, with Brazil, Paraguay, and Colombia as the largest producers of hydro-based electricity. However, thermal generation will also continue to represent an important share, mainly because of the preference for natural gas–based technologies, Mexico being the strongest example of that tendency. Among renewable sources other than hydropower, wind is estimated to grow at the fastest pace (16.2 percent annually). Still, it is not expected that wind-based generation will exceed more than 1 percent of the region's total generation by the end of the estimation period.

The regionwide implication of these shifts is that Latin America and the Caribbean will become slightly more carbon-intensive over the forecast period, driven mainly by the decline in the share of hydropower. However, it is difficult to make definitive conclusions about the future technological advances and market development of technologies in long-term estimation scenarios. For example, although wind power is expected to increase its share in the region's electricity generation mix by only a modest degree (from 0.1 percent to 1.3 percent of the total), many argue that, following the trajectory of wind power development in other regions, the share of this generation source in the electricity mix of Latin America and the Caribbean could grow to a much higher level than is currently estimated.

As noted earlier for the Latin America and the Caribbean region as a whole, the ICEPAC modeling exercise results in an estimate that an additional 239 GW of installed capacity will be required to satisfy the estimated demand. As shown in figure 3.3, it is estimated that Brazil will add approximately 41 percent of this additional capacity. The Southern Cone is expected to be the second-largest contributor, with about 45 GW, followed closely by Mexico and the Andean Zone, with about 44 GW and 30 GW, respectively.

Figure 3.4 provides a breakdown of the additional and total regionwide capacity by technology under the ICEPAC scenario. The middle column presents the incremental change (the delta). Table 3.5 depicts the percentage breakdown of capacity by technology to be added by 2030, showing that new generation will come mainly from hydropower and natural gas (36 percent and 35 percent, respectively). Under the ICEPAC scenario, the addition of hydropower capacity would bring its share in the 2030 regionwide total generation mix to nearly 50 percent. Coal would represent 11 percent of the added capacity. Finally, nuclear energy regionwide would comprise 4 percent of the added capacity, most of which would be contributed by Argentina.

One question that arises from the results of the ICEPAC scenario is whether there are alternatives to the current plans of Latin America and

Figure 3.3   Additional Capacity by Subregion

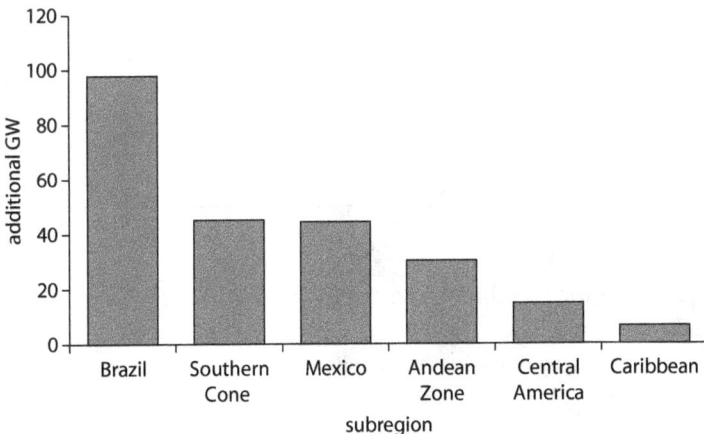

Source: Authors' elaboration based on optimization model.

Figure 3.4   Additional Capacity by Technology in Latin America and the Caribbean, 2008–30

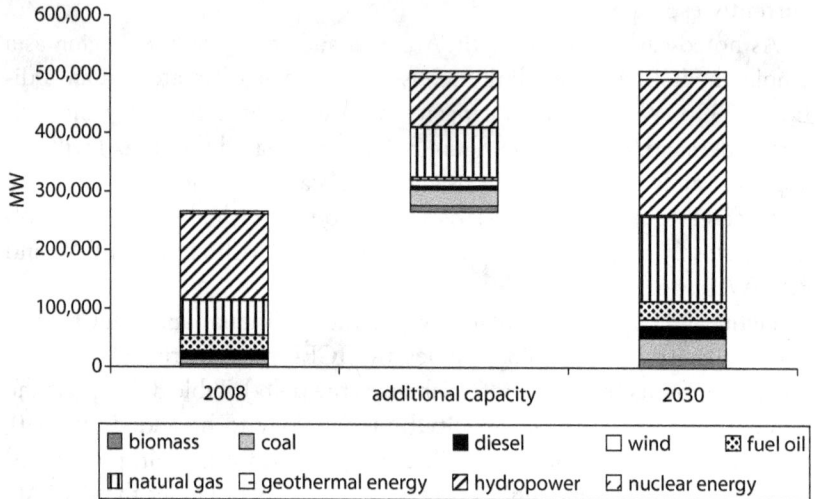

Source: Authors' elaboration based on optimization model.
Note: MW = megawatt.

Table 3.5   Additional Capacity by Technology in Latin America and the Caribbean, 2008–30

percent

Source	Additional capacity
Nuclear energy	4
Hydropower	36
Geothermal energy	0
Natural gas	35
Fuel oil	2
Wind	4
Diesel	3
Coal	11
Biomass	4

Source: Authors' elaboration.

the Caribbean for meeting its electricity needs that are feasible, environmentally sustainable, and efficient. Chapter 4 addresses some alternatives to achieving regionwide electricity security, including through more focused attention on renewable energy sources, increased regional integration and electricity trade, and improved supply- and demand-side energy efficiency. A second question is whether some of the plans for hydropower (and to some extent for natural gas) are realistic given the current policy environment in certain countries. To achieve the projected steep increase in hydropower between 2007 and 2030, the countries in the region likely will need to devote large, up-front investments to hydropower, more effectively attract private investment, and ambitiously pursue the development of untapped hydropower potential, while addressing the potentially negative environmental and social impacts of hydropower development.

Results for the subregions: The results of the ICEPAC scenario for the generation mix by subregion vary and are revealing in their similarities and differences. They also provide further insight into the implications and drivers of the energy mix over time.

Andean Zone: Under the ICEPAC scenario, the Andean Zone's generation demand is expected to grow 75 percent during the period of analysis. The generation mix is expected to be dominated by hydropower. Figure 3.5 illustrates how, in the short run, the mix holds constant, reflecting the fact that the current infrastructure is not being used to its maximum capacity. However, beginning in 2016, the share of fossil fuel–based generation is expected to increase, reaching 40 percent by 2030 (table 3.6). The subregion is not expected to diversify its generation mix by expanding into

Figure 3.5   Electricity Generation Mix in the Andean Zone, 2008–30

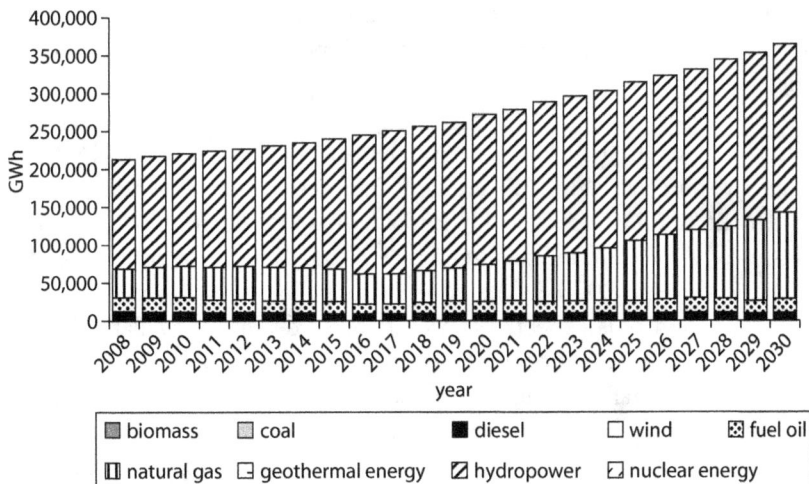

Source: Authors' elaboration based on optimization model.

Table 3.6   Electricity Generation Mix in the Andean Zone, 2008–30

percent

Source	Mix, 2008	Mix, 2030
Biomass	0	0
Coal	0	0
Diesel	6	3
Wind	0	0
Fuel oil	9	5
Natural gas	18	31
Geothermal energy	0	0
Hydropower	69	61
Nuclear energy	0	0

Source: Authors' elaboration.

nuclear, geothermal, or wind technologies. The implications for the Andean Zone are that the subregion will become increasingly dependent on fossil fuels (mainly natural gas) after 2016 and, as a result, will become more affected by volatile fossil fuel prices. However, as noted in appendix A, the projected trends vary within the subregion, with an expected increase in the share of hydropower in Ecuador.

Similar to the regionwide estimations, the additional capacity requirements from 2008 to 2030 in the Andean Zone under the ICEPAC scenario will be met largely by hydropower, with more than 60 percent of total capacity (40 percent of the additional generating capacity), and natural gas, with an estimated capacity of 31 percent (50 percent of the additional generating capacity), as illustrated in figure 3.6 and table 3.7. Diesel is estimated to reduce its share to 3 percent, while fuel oil is estimated to grow by 10 percent.

Brazil: Under the ICEPAC scenario, demand in Brazil increases by more than 160 percent. Figure 3.7 and table 3.8 present Brazil's generation mix under the ICEPAC scenario. Hydropower-based generation has prevailed historically in Brazil, and this trend is expected to hold under the ICEPAC scenario through 2030. In the latter part of the estimation period, however, the country is expected to experience a decline in available hydropower sources—a factor that will be compounded by a significant increase in overall electricity demand. Despite a continued increase in production from hydropower in absolute terms, the aforementioned factors would necessitate an increased reliance on natural gas as well as on coal. Additionally, the country's plans include the completion of the Angra 3 nuclear power plant.

Figure 3.6   Additional Capacity by Technology in the Andean Zone, 2008–30

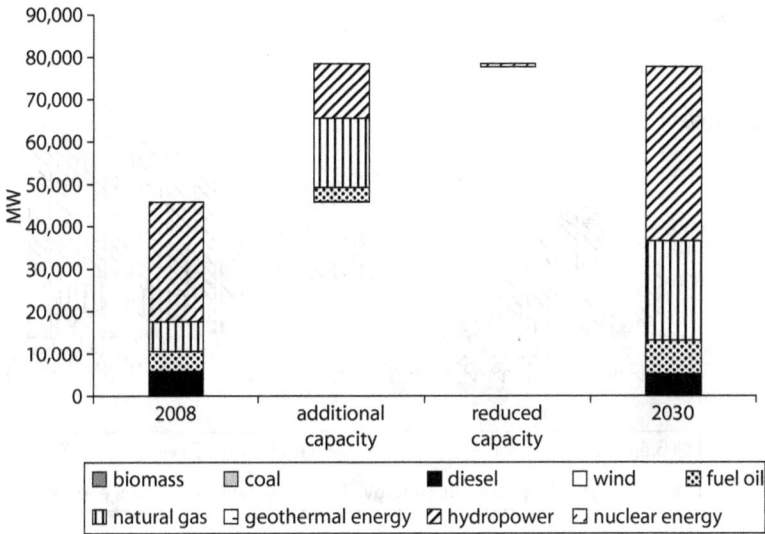

Source: Authors' elaboration based on optimization model.
Note: MW = megawatt.

Table 3.7   Additional Capacity by Technology in the Andean Zone, 2008–30

percent

Source	Additional capacity
Nuclear energy	0
Hydropower	40
Geothermal energy	0
Natural gas	50
Fuel oil	10
Wind	0
Diesel	0
Coal	0
Biomass	0

Source: Authors' elaboration.

For Brazil, the ICEPAC scenario indicates that hydropower will remain the dominant source in the electricity generation mix and, as indicated in figure 3.8 and table 3.9, will represent the largest share of additional capacity—45 percent. However, although the overall increase in hydropower capacity will be substantial in absolute terms, the share of hydropower in the electricity matrix will actually decline, from

Figure 3.7   Electricity Generation Mix in Brazil, 2008–30

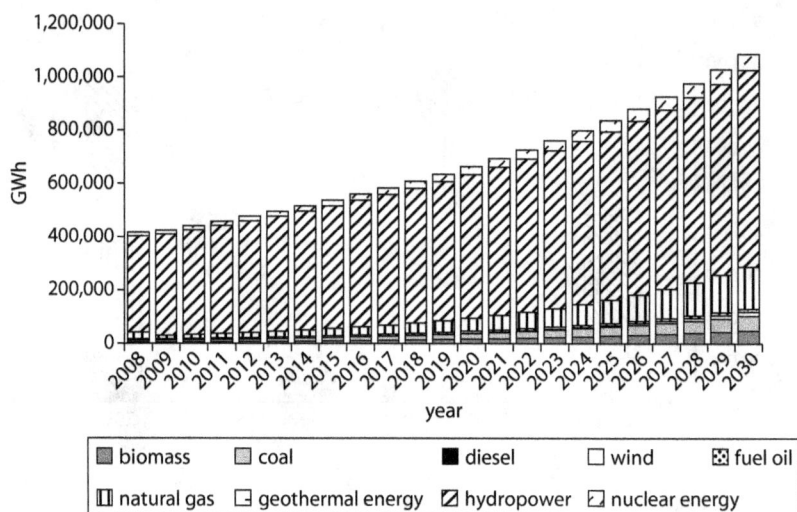

Source: Authors' elaboration based on optimization model.

Table 3.8   Electricity Generation Mix in Brazil, 2008–30
percent

Source	Mix, 2008	Mix, 2030
Biomass	1	4
Coal	1	5
Diesel	1	0
Wind	0	2
Fuel oil	1	1
Natural gas	6	15
Geothermal energy	0	0
Hydropower	87	68
Nuclear energy	3	5

Source: Authors' elaboration.

87 percent in 2008 to 68 percent in 2030. Brazil's additional hydropower capacity is still expected to be the dominant contributor to the regionwide increase in the share of hydropower, whereas the Southern Cone and the Andean Zone are forecast to follow with 12 GW and 13 GW, respectively. It is also important to mention that the projected increase in biomass-based generation, largely from sugarcane bagasse, will be from about 1 percent to 4 percent between 2008 and 2030.

Figure 3.8   Additional Capacity by Technology in Brazil, 2008–30

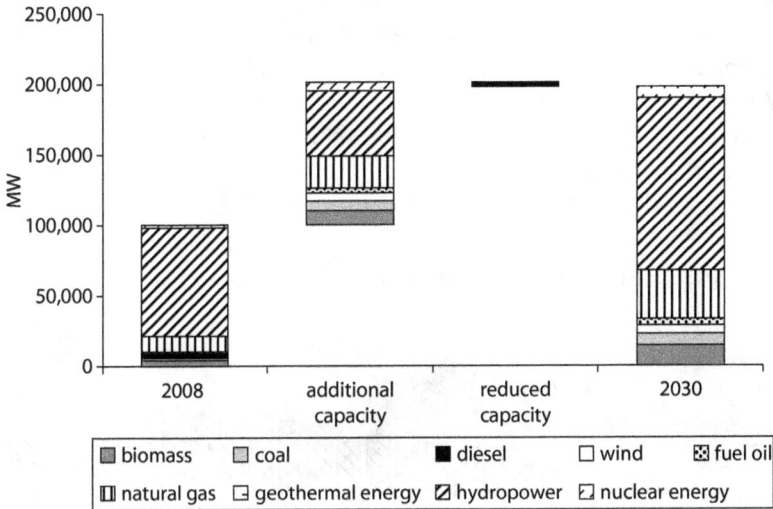

Source: Authors' elaboration based on optimization model.
Note: MW = megawatt.

Table 3.9   Additional Capacity by Technology in Brazil, 2008–30

percent

Source	Additional capacity
Nuclear energy	6
Hydropower	45
Geothermal energy	0
Natural gas	23
Fuel oil	3
Wind	6
Diesel	0
Coal	7
Biomass	10

Source: Authors' elaboration.

Southern Cone: Under the ICEPAC scenario, demand in the Southern Cone will grow nearly 80 percent. Figure 3.9 shows the Southern Cone's generation mix. The subregion's existing high degree of diversification of generation technologies and fuels will become even more dynamic over the period from 2008 to 2030. The Southern Cone's generation capacity is projected to be dominated by hydropower (45 percent) and natural gas (29 percent) as illustrated in table 3.10. The fuel source that most increases in the ICEPAC scenario is coal, which rises from 7 percent to 12 percent, largely as a result of the cost-minimization assumptions in the model (which do not include global environmental costs). Renewable sources, including wind and geothermal energy, are expected to increase modestly, with their share rising from close to 0 percent to 3 percent. This increase is driven in part by an anticipated increase in the Argentinean wind market. It is important to note that

Figure 3.9   Electricity Generation Mix in the Southern Cone, 2008–30

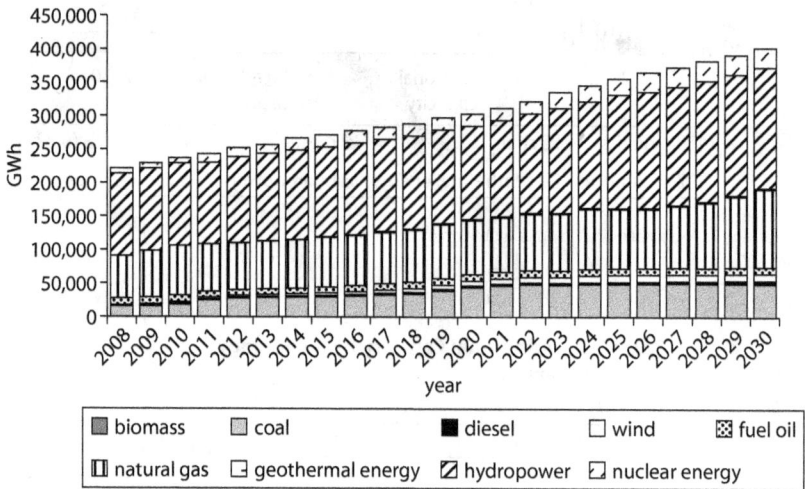

Source: Authors' elaboration based on optimization model.

Table 3.10   Electricity Generation Mix in the Southern Cone, 2008–30

percent

Source	Mix, 2008	Mix, 2030
Biomass	0	0
Coal	7	12
Diesel	1	1
Wind	0	3
Fuel oil	5	2
Natural gas	28	29
Geothermal energy	0	0
Hydropower	56	45
Nuclear energy	3	7

Source: Authors' elaboration.

the increases in nuclear, coal, and non-hydro renewables will take place at the expense of hydropower, whose share is estimated to decline. Nuclear energy accounts for 6 percent of additional capacity as a result of new capacity planned in Argentina. Nonetheless, as illustrated in figure 3.10 and table 3.11, the Southern Cone's additional capacity is projected to be dominated by natural gas (33 percent) and hydropower (28 percent).

Figure 3.10   Additional Capacity by Technology in the Southern Cone, 2008–30

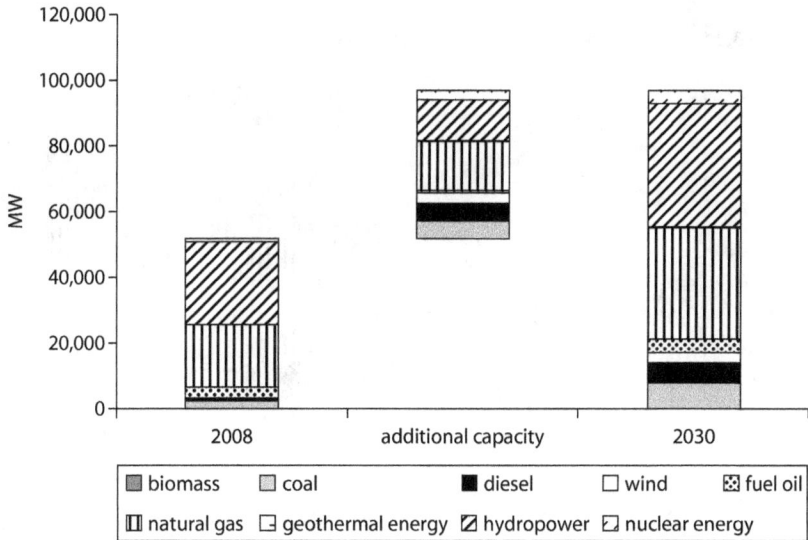

Source: Authors' elaboration based on optimization model.
Note: MW = megawatt.

Table 3.11   Additional Capacity by Technology in the Southern Cone, 2008–30

percent

Source	Mix, 2008
Nuclear energy	6
Hydropower	28
Geothermal energy	0
Natural gas	33
Fuel oil	2
Wind	7
Diesel	12
Coal	12
Biomass	0

Source: Authors' elaboration.

Mexico: According to Mexico's official development plan, the bulk of the additional generation capacity will be based on integrated gasification combined-cycle technologies. Given the country's natural resource base, natural gas technologies will remain the dominant source of electricity in both the short and the long term (figure 3.11). Under Mexico's baseline plan, the share of natural gas remains relatively constant through 2014, and then increases rapidly. The share of natural gas in the electricity generation mix rises by 15 percent (from 50 percent to 65 percent), with a parallel decline in the share of fuel oil (table 3.12). The share of coal

Figure 3.11   Electricity Generation Mix in Mexico, 2008–30

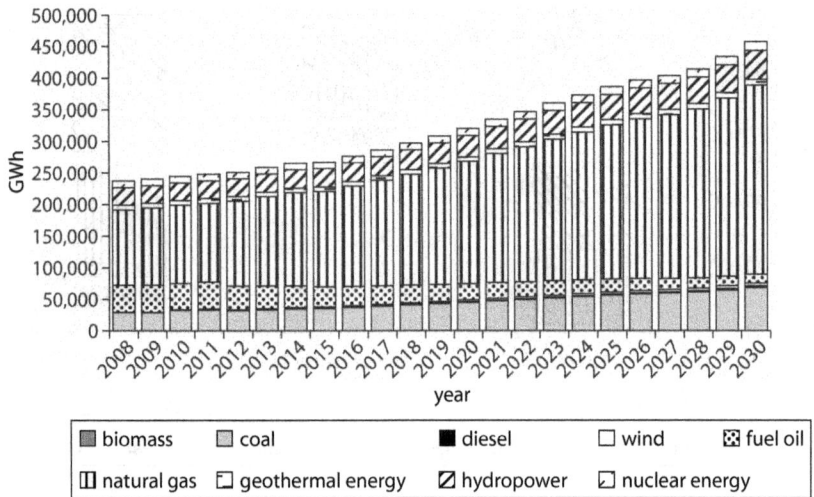

Source: Authors' elaboration based on optimization model.

Table 3.12   Electricity Generation Mix in Mexico, 2008–30
percent

Source	Mix, 2008	Mix, 2030
Biomass	0	0
Coal	12	15
Diesel	0	1
Wind	0	1
Fuel oil	18	3
Natural gas	50	65
Geothermal energy	3	2
Hydropower	12	10
Nuclear energy	4	3

Source: Authors' elaboration.

increases marginally, from 12 percent to 15 percent. The shares of other technologies remain largely proportionate to their 2008 levels (figure 3.12 and table 3.13).

Central America: For Central America, under the ICEPAC scenario the share of natural gas in the electricity generation mix increases dramatically

Figure 3.12 Additional Capacity by Technology in Mexico, 2008–30

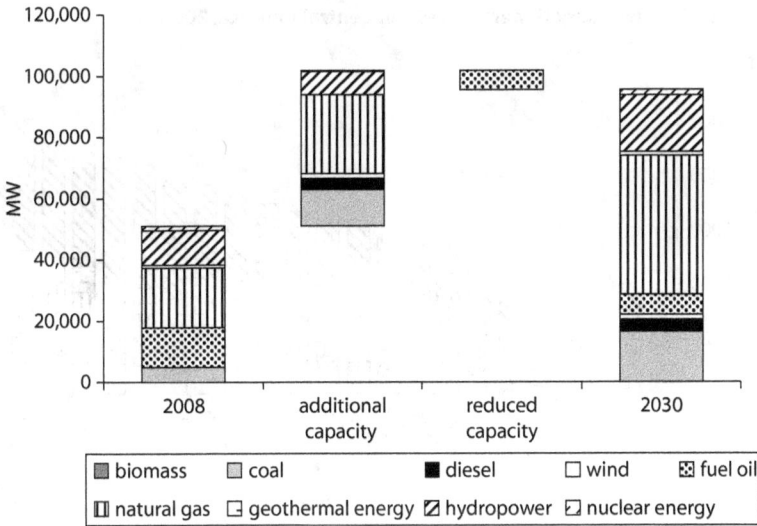

Source: Authors' elaboration based on optimization model.
Note: MW = megawatt.

Table 3.13   Additional Capacity by Technology in Mexico, 2008–30

percent

Source	Additional capacity
Nuclear energy	1
Hydropower	14
Geothermal energy	0
Natural gas	51
Fuel oil	0
Wind	3
Diesel	8
Coal	23
Biomass	0

Source: Authors' elaboration.

between 2008 and 2030 from 0 percent to 13 percent (figure 3.13 and table 3.14). Central America is also the only subregion where the share of hydropower marginally increases between 2008 and 2030 (from 45 percent to 46 percent of total generation). Similarly, the subregion is expected to increase its reliance on coal, from about 2 percent in 2008 to more than 11 percent by 2030. By contrast, the share of fuel oil and diesel is expected to decline significantly, largely because of the preference to diversify away from fossil fuels and the expected changes in relative generation costs.

Figure 3.13   Electricity Generation Mix in Central America, 2008–30

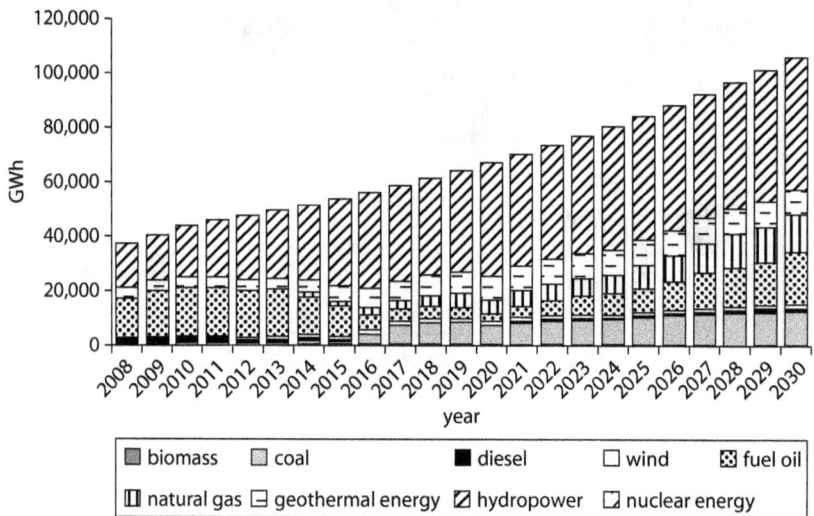

Source: Authors' elaboration based on optimization model.

Table 3.14   Electricity Generation Mix in Central America, 2008–30

percent

Source	Mix, 2008	Mix, 2030
Biomass	0	0
Coal	2	11
Diesel	5	1
Wind	2	1
Fuel oil	35	18
Natural gas	0	13
Geothermal energy	11	9
Hydropower	45	46
Nuclear energy	0	0

Source: Authors' elaboration.

As shown in figure 3.14 and table 3.15, the bulk of generating capacity to be added in Central America between 2008 and 2030 is projected to be largely hydropower (45 percent), with coal, fuel oil, and natural gas also playing important roles (11 percent, 20 percent, and 14 percent, respectively, of additional capacity). Renewable energy sources, such as wind and geothermal energy, begin to play an increasingly important role

Figure 3.14   Additional Capacity by Technology in Central America, 2008–30

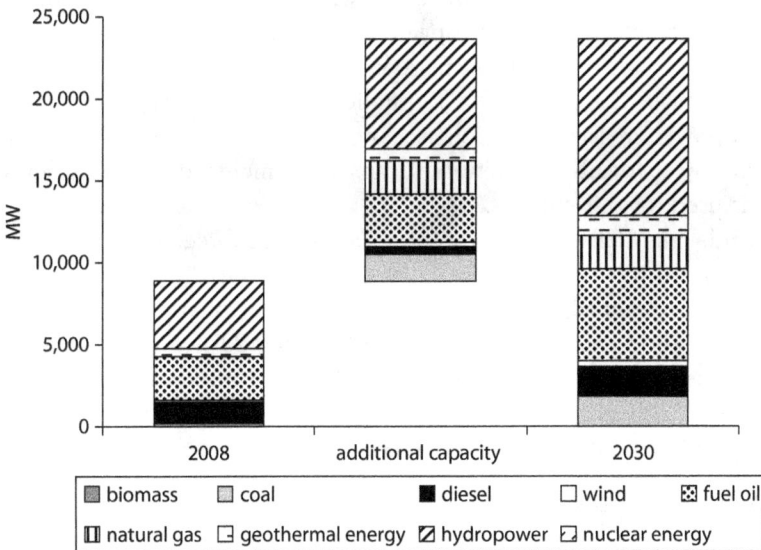

Source: Authors' elaboration based on optimization model.
Note: MW = megawatt.

Table 3.15   Additional Capacity by Technology in Central America, 2008–30

percent

Source	Additional capacity
Nuclear energy	1
Hydropower	45
Geothermal energy	5
Natural gas	13
Fuel oil	20
Wind	2
Diesel	4
Coal	11
Biomass	0

Source: Authors' elaboration.

in Central America under the ICEPAC scenario, and together represent about 7 percent of the new generating capacity installed by 2030. The potential and prospects for wind power in the subregion are analyzed in more detail in chapter 4.

The Caribbean: In the Caribbean, the generation matrix is expected to remain largely fossil fuel–dependent (figure 3.15). Over the entire period, the share of fossil fuel–based technologies increases slightly from 93 percent to 95 percent (table 3.16) with fuel oil and diesel contributing 32 percent of the total capacity in 2030. One important development under the ICEPAC scenario is an increase in natural gas–based generation in the Dominican Republic. By contrast, the proportion of hydropower in the subregion is expected to decrease: No additional hydropower (or other renewable) capacity is included in the expansion plans used for this analysis.

Under the ICEPAC scenario, the Caribbean subregion is expected to continue to rely largely on conventional electricity generation sources, with fuel oil and diesel contributing 28 percent of the added capacity between 2008 and 2030 and coal contributing 23 percent (figure 3.16 and table 3.17). The bulk of future additional capacity (43 percent) in

Figure 3.15   Electricity Generation Mix in the Caribbean, 2008–30

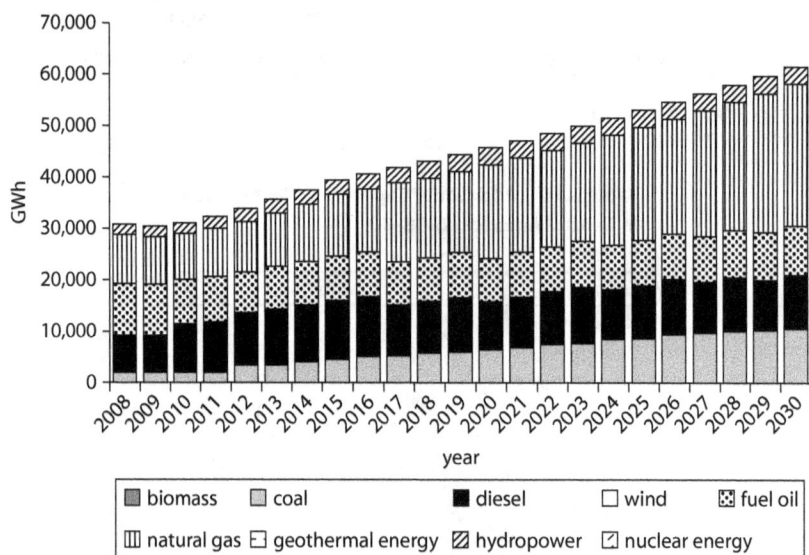

Source: Authors' elaboration based on optimization model.

Table 3.16   Electricity Generation Mix in the Caribbean, 2008–30

percent

Source	Mix, 2008	Mix, 2030
Biomass	0	0
Coal	6	17
Diesel	24	17
Wind	0	0
Fuel oil	32	15
Natural gas	31	45
Geothermal energy	0	0
Hydropower	7	5
Nuclear energy	0	0

Source: Authors' elaboration.

Figure 3.16   Additional Capacity by Technology in the Caribbean, 2008–30

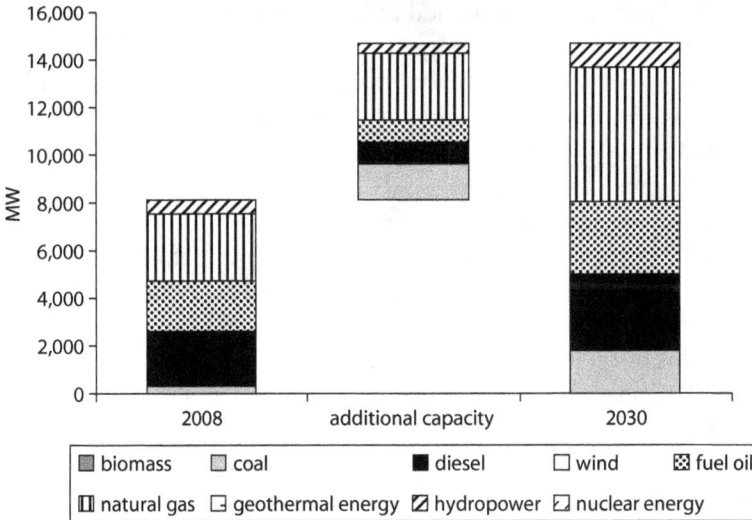

Source: Authors' elaboration based on optimization model.
Note: MW = megawatt.

this scenario is based on natural gas technologies, driven mainly by the Dominican Republic.

Implications for Investment Needs

Under the ICEPAC scenario, the total investment for additional electricity generation infrastructure in Latin America and the Caribbean reaches

Table 3.17   Additional Capacity by Technology in the Caribbean, 2008–30

percent

Source	Additional capacity
Nuclear energy	0
Hydropower	6
Geothermal energy	0
Natural gas	43
Fuel oil	14
Wind	0
Diesel	14
Coal	23
Biomass	0

Source: Authors' elaboration.

US$430 billion during the period 2008–30. This total is estimated using the additional capacity requirements for each subregion and the assumed fixed costs per installed megawatt for each technology (see appendix C for cost assumptions). This amount does not include variable and fuel costs and would be higher when considering the necessary investments for additional transmission and distribution infrastructure. The distribution of investment requirements by subregion is as follows: Brazil—42 percent, Southern Cone—18 percent, Mexico—18 percent, the Andean Zone—13 percent, Central America—6 percent, and the Caribbean—2 percent (figure 3.17).

In terms of the required investment as a share of GDP (figure 3.18), the ICEPAC exercise indicates that Central America and Brazil will have the highest spending requirements, with each requiring an average of 0.6 percent of their GDP per year. This cost estimation does not include fuel cost.[7] The Southern Cone and the Andean Zone would need additional investment amounting to 0.4 percent of GDP per year on average. The Caribbean is projected to require additional investments equivalent to 0.28 percent of GDP. Mexico has the lowest additional investment requirements as a share of GDP (0.2 percent), mainly as a consequence of its current excess capacity.

Implications for CO_2 Emissions

With the ICEPAC scenario's generation mix by technology, CO_2 emissions from the power sector in Latin America and the Caribbean would

Figure 3.17   Share of Investment Requirements by Subregion, 2008–30

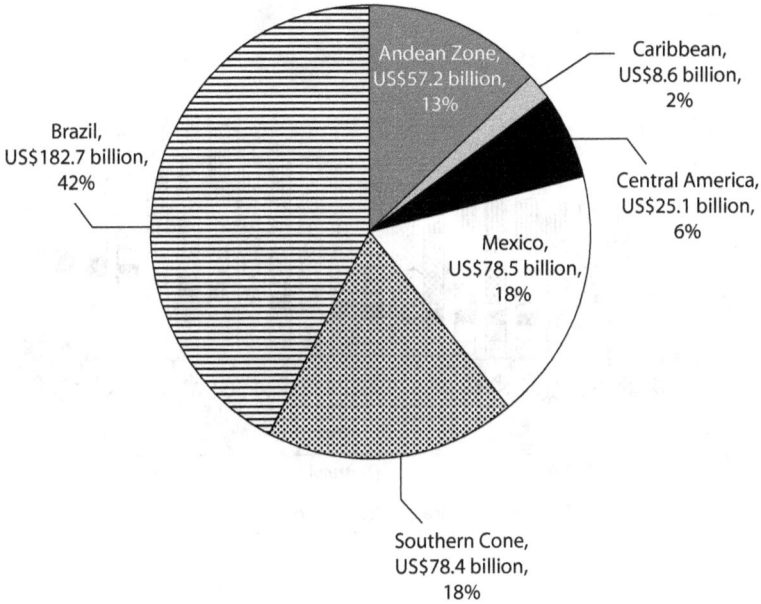

Source: Authors' elaboration based on optimization model.

Figure 3.18   Average Investment Requirement in Electricity Generation Capacity as a Share of GDP

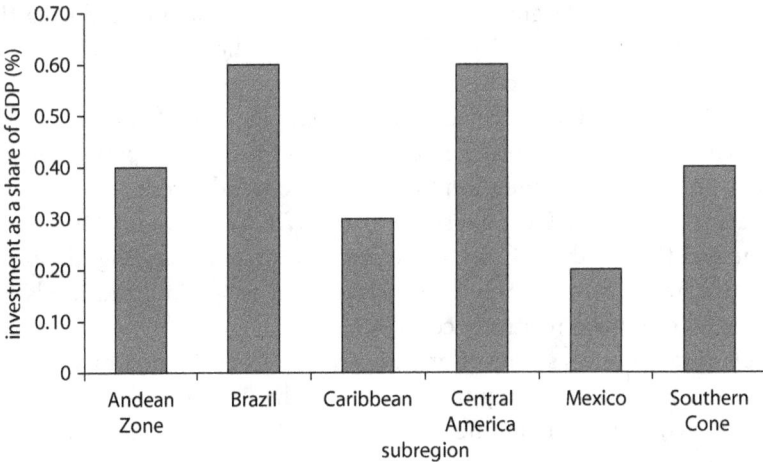

Source: Authors' elaboration based on optimization model.

Figure 3.19   ICEPAC Scenario CO_2 Emissions, Latin America and the Caribbean

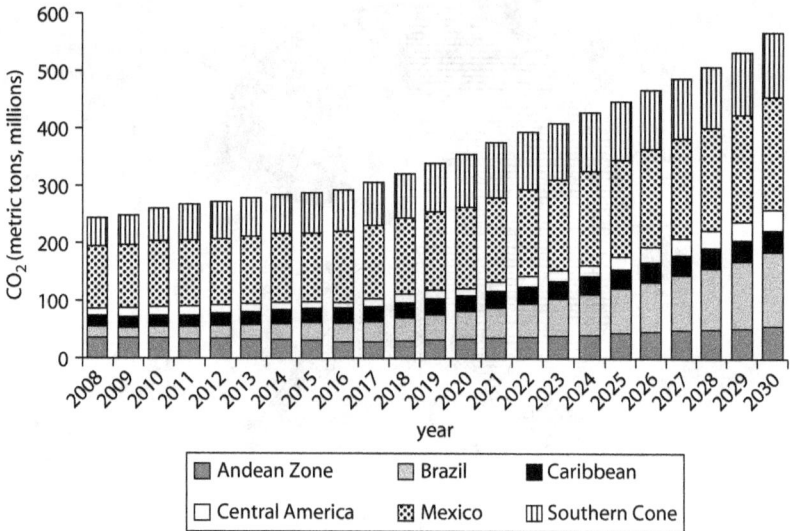

Source: Authors' elaboration based on optimization model.

more than double by 2030 (figure 3.19). In 2008, 244 million metric tons of CO_2 were emitted into the atmosphere; by 2030, emissions would reach nearly 569 million metric tons of CO_2.[8]

The ICEPAC scenario shows that Brazil and Mexico would emit more CO_2 per generated unit of electricity (GWh) in 2030 than in 2008 (figure 3.20). Brazil's gradual increase in emissions per GWh during the latter part of the period 2008–30 is driven by the diversification of its energy matrix away from hydropower.

Coal consumption is projected to increase in Central America under the ICEPAC scenario; however, its reliance on fuel oil for power generation would decrease. When combined, the two trends produce a relatively constant amount of CO_2 emissions per GWh. Similarly in the Andean Zone and the Southern Cone, emissions levels would vary slightly in the medium term; however, over the entire forecast period, the levels are expected to remain relatively constant.

Despite Mexico's strong efforts to diversify its generation mix by adding more renewable energy to the grid, the hydrocarbon sector (mainly natural gas) would remain the dominant fuel for power generation under the ICEPAC scenario. In addition, the country's overall emissions intensity would increase slightly.

Figure 3.20   ICEPAC Scenario CO$_2$ Emissions by Subregion

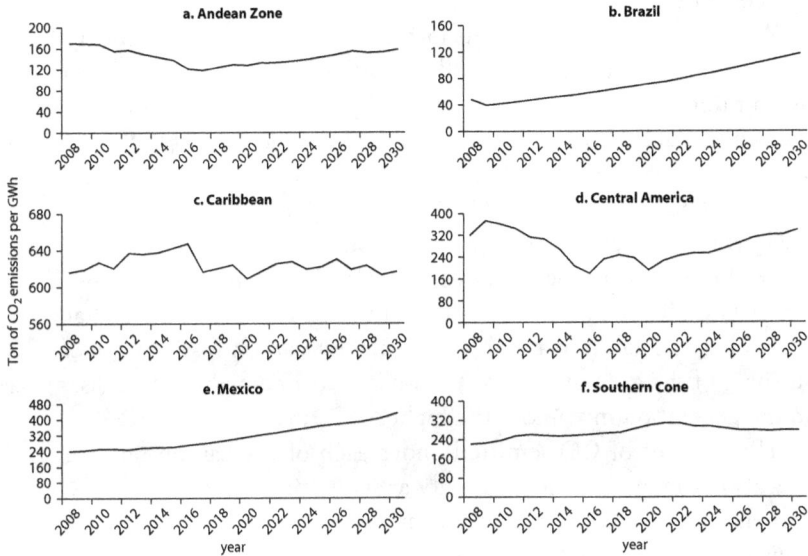

Source: Authors' elaboration based on optimization model.

It is worth repeating that the purpose of the long-run ICEPAC exercise is to provide a view of the overall regional and subregional trends in supply and demand, rather than to provide specific numbers for individual countries and individual technologies. The results are based on the assumptions, limitations, and constraints of the modeling exercise. A comparison between the ICEPAC assumptions and those of the latest country expansion plans is provided in appendix D.

Carbon Taxes: Illustrative Effect on Choice among Fossil Fuels

Limited sensitivity analysis was undertaken with the model to show the effect of a carbon tax. The sensitivity analysis was limited in the sense that, effectively, only electricity generation using fossil fuels was allowed to vary with the imposition of carbon taxes. In the model, the projected share of renewables in the generation mix is based on the maximum levels projected in the countries' own expansion plans and, thus, is treated as a fixed input into the model. As a result, the sensitivity analysis is a fossil fuel–based generation sensitivity analysis only, rather than a sensitivity analysis inclusive of all technologies.[9] Because the model assumes that the amount of natural gas, coal, fuel oil, and diesel is unconstrained, the carbon tax does have an effect on how much electricity from these sources is demanded.

With these important caveats in mind, we considered the effect of the introduction of carbon taxes on the choice of fossil fuels. As noted in the assumptions about price, the baseline oil price used in the modeling exercise was a real price of US$100 between 2008 and 2030. For the sensitivity analysis, two carbon tax scenarios were evaluated: US$20 per ton of CO_2 and US$50 per ton of CO_2. The conclusion from the analysis (figure 3.21) is that a CO_2 tax of US$20 per ton has no effect on the selection of fossil fuel–based generation technologies. However, with a higher CO_2 tax of US$50 per ton, there is sufficient motivation to cause a change in technology selection. Natural gas, the lowest-carbon fossil fuel, increases its share of the generation mix. Coal-based technologies almost disappear in the generation mix under the higher CO_2 tax.

The amount of CO_2 emitted under each of the carbon tax scenarios was also estimated in the sensitivity analysis (figure 3.22). As expected, a carbon tax does reduce projected carbon emissions. In the scenario with a high carbon tax (US$50 per ton), emissions are as much as 30 percent lower than in the scenario with no carbon tax. The effect of the carbon

Figure 3.21   Sensitivity Analysis of CO_2 Tax and Composition Generation Mix, Fossil Fuels Only

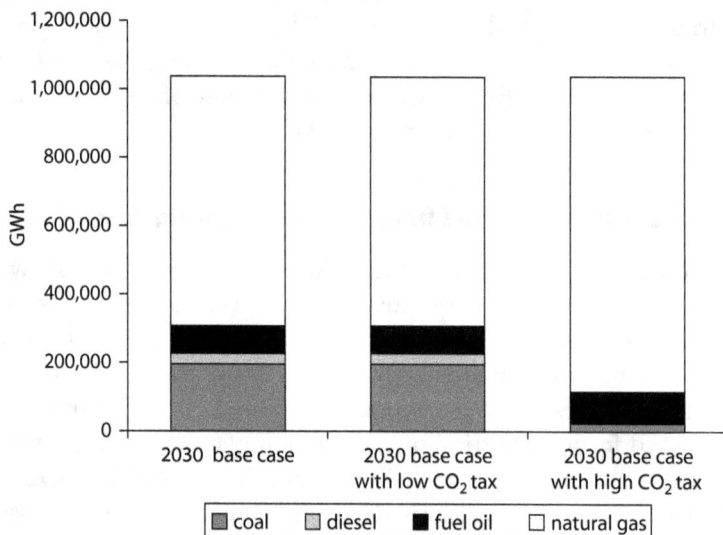

Source: Authors' elaboration based on optimization model.

Figure 3.22   Sensitivity Analysis of Greenhouse Gas Emissions in Latin America and the Caribbean

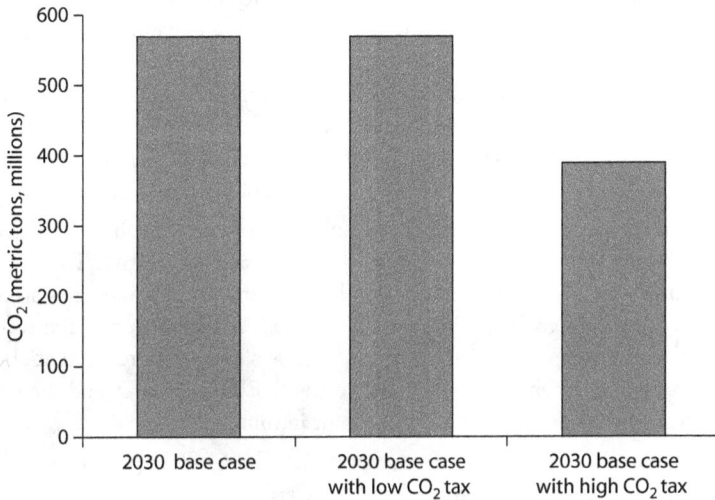

Source: Authors' elaboration based on optimization model.

tax would be expected to be considerably larger if the model allowed for the substitution of renewables for fossil fuels, as opposed to simply switching between fossil fuels of differing carbon intensity.

Notes

1. The "electricity price" variable refers to each country's weighted average electricity tariff according to OLADE's database.

2. Electricity demand models typically are estimated using economic growth and electricity prices.

3. Information from the International Monetary Fund on GDP and OLADE electricity prices for countries across the region was used.

4. We initially used an econometric specification in which electricity demand is jointly driven by GDP, population, and electricity price. However, high multicollinearity between GDP and population growth produced artificially inflated standard errors and R^2. Although multicollinearity does not violate any of the standard model's assumptions, the nature of this exercise requires the highest possible precision on the coefficient estimates, particularly the GDP elasticity. As such, the preferred model includes GDP and electricity prices as explanatory variables.

5. Relying on both practical and theoretical considerations, the team considered it was reasonable to hold electricity prices constant throughout the projection period. All else being equal, the GDP elasticity estimate characterizes the partial effect of economic growth on electricity demand, independent of changes in prices for fuels used in electricity generation or in capital costs. Furthermore, because electricity prices are endogenous, a full supply-demand system of equations would need to be estimated, which is beyond the purpose of this exercise. Moreover, such specification would require precise assumptions (and forecasts) about the fuels and technologies used in generation.

6. For the years in which no expansion plan data were available, the ICEPAC scenario uses thermal generation as the expansion alternative. The decision was made to rely on this assumption because thermal technologies are considered tradable goods in the market, whereas renewable generation sources, although not fixed, are more determined by available resource potential. This assumption is a significant constraint to the ICEPAC scenario, and the results should be considered with this constraint in mind.

7. In the past decade, the cost of oil imports as a share of GDP has shown a slow but steady rise for Latin America and the Caribbean as well as for the Central American and Caribbean subregions. The cost of oil imports in Latin America and the Caribbean increased from 1 percent of GDP in 1998 to 4 percent in 2008. However, for Central America and the Caribbean the cost of oil imports in 2008 was 6 percent and 8 percent, respectively. Furthermore, there are specific countries within Central America and the Caribbean where the oil share of GDP was 10 percent or higher in 2008: Guyana at 20 percent, Jamaica at 16 percent, and Nicaragua at 10 percent.

8. This amount is the authors' estimation considering the power generated by technology in 2008.

9. This distinction under the model is most evident when viewing the share of hydropower; namely, the hydropower share in each country's generation mix, similar to those of nuclear, wind, geothermal, and biomass energy, remains constant irrespective of the oil price and carbon tax, because the baseline scenario already uses the maximum hydropower potential. As carbon taxes are imposed, hydropower becomes relatively cheaper, but the share under the model does not increase because the resource is limited.

References

Acton, J., B. Mitchell, and R. Mowill. 1976. "Residential Demand for Electricity in Los Angeles: An Econometric Study of Disaggregated Data." Report R-1899-NSF, The Rand Corporation, Santa Monica, CA.

Anderson, K. P. 1973. "Residential Demand for Electricity: Econometric Estimates for California and the United States." *Journal of Business* 46 (4): 526–53.

Barnes, R., R. Gillingham, and R. Hagemann. 1981. "The Short-Run Residential Demand for Electricity." *Review of Econometrics and Statistics* 63 (4): 541–51.

Beenstock, M., E. Goldin, and D. Nabot. 1999. "The Demand for Electricity in Israel." *Energy Economics* 21 (2): 168–83.

Benavente, J.M., A. Galetovic, R. Sanhueza, and P. Serra. 2005. "Estimando la Demanda Residencial por Electricidad en Chile: El Consumo es Sensible al Precio." Cuadernos de Economía, 42 (mayo): 31–61.

Bjørner, T. B., and H. H. Jensen. 2002. "Interfuel Substitution within Industrial Companies: An Analysis Based on Panel Data at Company Level." *Energy Journal* 23 (2): 27–50.

Boonekamp, P. G. M. 2007. "Price Elasticities, Policy Measures and Actual Developments in Household Energy Consumption—A Bottom Up Analysis for the Netherlands." *Energy Economics* 29 (2): 133–57.

Brännlund, R., T. Ghalwash, and J. Nordström. 2007. "Increased Energy Efficiency and the Rebound Effect: Effects on Consumption and Emissions." *Energy Economics* 29 (1): 1–17.

Caloghirou, Y. D., A. G. Mourelatos, and H. Thompson. 1997. "Industrial Energy Substitution during the 1980s in the Greek Economy." *Energy Economics* 19 (4): 476–91.

Chang, Y., and E. Martinez-Combo, 2003. "Electricity Demand Analysis Using Cointegrating and Error-Correction Models with Time Varying Parameters: The Mexican Case." Working Paper 2003-08, Department of Economics, Rice University, Houston.

Espey, J., and M. Espey. 2004. "Turning on the Lights: A Meta-Analysis of Residential Electricity Demand Elasticities." *Journal of Agricultural and Applied Economics* 36 (1): 65–81.

Filippini, M., and S. Pachuari. 2002. "Elasticities of Electricity Demand in Urban Indian Households." CEPE Working Paper 16, Centre for Energy Policy and Economics, Swiss Federal Institutes of Technology, Zurich.

Fisher, F. M., and C. Kaysen. 1962. *A Study in Econometrics: The Demand for Electricity in the United States.* Amsterdam: North-Holland.

Galetovic, A., R. Sanhueza, and P. Serra. 2001. Estimacion de los costos de falla residencial y comerical. Draft. Santiago, Chile.

Halvorsen, R. 1975. "Residential Demand for Electric Energy." *Review of Economics and Statistics* 57 (1): 12–18.

Holtedahl, P., and F. L. Joutz. 2004. "Residential Electricity Demand in Taiwan." *Energy Economics* 26 (2): 201–24.

Houthakker, H. 1980. "Residential Electricity Revisited." *Energy Journal* 1 (1): 29–42.

Houthakker, H. S., and L. D. Taylor. 1970. *Consumer Demand in the United States*, 2nd ed. Cambridge, MA: Harvard University Press.

Houthakker, H. S., P. K. Verleger, and D. P. Sheehan. 1973. "Dynamic Demand Analyses for Gasoline and Residential Electricity." *American Journal of Agricultural Economics*, 56 (2): 412–18.

IMF (International Monetary Fund). 2009. World Economic Outlook Database (April 2009). World Bank, Washington, DC. http://www.imf.org/external/pubs/ft/weo/2009/01/weodata/index.aspx.

Irffi, G., I. Castelar, M. Siqueira, and F. Linhares. 2006. "Dynamic OLS and Regime Switching Models to Forecast the Demand for Electricity in the Northeast of Brazil." Graduate School of Economic, Getulio Vargas Foundation, Brazil.

Maddock, R., E. Castano, and F. Vella. 1992. "Estimating Electricity Demand: The Cost of Linearising the Budget Constraint." *Review of Economics and Statistics* 74 (2): 350–54.

McFadden, D., C. Puig, and D. Kirshner. 1977. "A Simulation Model for Electricity Demand." Final Report, Cambridge Systematics, Inc., Cambridge, MA.

Modiano. E. M. 1984. "Elasticidade-renda e preço da demanda de energia elétrica no Brasil." Rio de Janeiro: PUC/RJ (Texto para discussão, 68).

Mount, T., L. Chapman, and T. Tyrrell, 1973. "Electricity Demand in the United States: An Econometric Analysis." ORNL-NSF-EP-49, Oak Ridge National Laboratories, Tennessee, USA.

Murray, M., R. Spann, L. Pulley, and E. Beauvais. 1978. "The Demand for Electricity in Virginia." *Review of Economics and Statistics* 60 (4): 585–600.

Roy, J., A. H. Sanstad, J. A. Sathaye, and R. Khaddaria. 2006. "Substitution and Price Elasticity Estimates Using Inter-Country Pooled Data in a Translog Cost Model." *Energy Economics* 28 (5–6): 706–19.

Schmidt, C. A., and M. A. M. Lima. 2004. "A demanda por energía elétrica no Brasil." *Revista Brasileira de Economia* 58 (1): 67–98.

Shin, J.-S. 1985. "Perception of Price When Price Information Is Costly: Evidence from Residential Electricity Demand." *Review of Economics and Statistics* 67 (4): 591–98.

Silk, J. I., and F. L. Joutz. 1997. "Short and Long-Run Elasticities in U.S. Residential Electricity Demand: A Co-integration Approach." *Energy Economics* 19 (4): 493–513.

Taylor, L. D., G. R. Blattenberger, and P. K. Verleger Jr. 1977. "The Residential Demand for Energy." Final Report, EPRI EA-235, vol. 1, Electric Power Research Institute, Palo Alto, CA.

Urga, G., and C. Walters. 2003. "Dynamic Translog and Linear Logit Models: A Factor Demand Analysis of Interfuel Substitution in U.S. Industrial Energy Demand." *Energy Economics* 25 (1): 1–21.

Wilder, R. P., and J. F. Willenborg. 1975. "Residential Demand for Electricity: A Consumer Panel Approach." *Southern Economic Journal* 42 (2): 212–17.

Wolak, F. A., and R. H. Patrick. 2001. "Estimating the Customer-Level Demand for Electricity under Real Time Market Prices." NBER Working Paper 8213, National Bureau of Economic Research, Cambridge, MA.

Zachariadis, T., and N. Pashourtidou. 2007. "An Empirical Analysis of Electricity Consumption in Cyprus." *Energy Economics* 29 (2): 183–98.

CHAPTER 4

Alternatives for Meeting Future Electricity Needs

Given the electricity demand and supply picture presented in chapter 3, it is clear that the Latin America and the Caribbean region will be challenged in meeting its future electricity needs. As noted previously, however, the electricity expansion plans for most countries in the region (and, thus, the basis of the modeling exercise in chapter 3) fail to fully account for a number of important options that could increase supply and lower demand. The purpose of this chapter is to briefly outline several of these options. The first section examines the potential for renewable energy sources, including not only an assessment of hydropower resources, which are already a large and important component of current expansion plans, but also potential contributions from non-hydro renewables for power generation, including biomass, geothermal energy, and wind. The second section examines the potential and benefits of greater cross-border trade within Latin America and the Caribbean, an option that could also help to catalyze a greater share of the region's renewable energy potential. The third section discusses the potential for supply-side and demand-side energy-efficiency measures.

Renewable Energy Potential

In 2007, the share of renewable energy in Latin America and the Caribbean was about 60 percent of the total power generation, the highest percentage of any region in the world. Of this, 57 percent was from hydropower and only 3 percent was from other renewable energy sources. As seen in chapter 3, the existing national expansion plans are quite aggressive with respect to large hydroelectric projects, but are very modest regarding small hydro and non-hydro renewables. Nonetheless, there are reasons to believe that small hydro and non-hydro renewables could be expanded much more than is envisioned under current country power expansion plans. The region has large renewable energy resources, ranging from wind in Argentina to hydroelectricity and biomass in Brazil[1] to geothermal energy in Central America, and solar is ubiquitous.[2] Climate-change concerns are spurring renewable energy development in many parts of the world, including Latin America, and countries are finding that there are other important benefits of renewable energy, including making use of local resources, energy diversification, and cleaner energy production.

The high proportion of renewable energy in Latin America and the Caribbean, mainly hydropower, has been falling gradually over the past two decades and is projected to decline even more rapidly through 2030. It is illustrative to consider what changes in the future power generation mix would be needed to halt this decline and to maintain the current share of renewable energy through 2030. Two conclusions seem clear: (a) the region must also work hard to achieve the aggressive plans for new hydropower plants, and (b) the use of renewable resources other than hydropower must expand significantly more than is currently projected in country expansion plans.

Hydropower

It is estimated that between 700 and 1,400 terawatt-hours (TWh) of economically viable hydropower potential remains to be developed in the Latin America and the Caribbean region. The wide range of values reflects the incomplete information available about inventories of hydroelectricity potential as well as the inherent uncertainties in development, which are discussed in the next section. Under the most optimistic estimates, there is enough hydroelectricity to supply the entire region's projected power expansion needs to 2030. Under the less optimistic hydroelectricity scenario, there would still be enough to supply about two-thirds of the expansion.

However, it is highly unlikely that the remaining hydropower potential in Latin America would be developed within the next 20–25 years. Hydropower resources are unevenly distributed—many in remote areas—and typically do not match the distribution of demand centers. In addition, hydropower takes time to develop. It is not unusual for a hydropower plant to span 10 years from planning to production. Brazil, which has about 40 percent of the region's hydropower resources and currently produces more than half the region's total, plans to develop almost all of its exploitable potential by 2030. Outside Brazil, however, the situation is quite different. For example, three countries—Colombia, Ecuador, and Peru—possess more than half of the total remaining potential but have developed less than 10 percent. Unless the hydropower potential of these countries can be integrated into the regional grid, it is unlikely that significant portions of that potential can be developed before 2030 given insufficient domestic demand.

In the supply scenarios presented in chapter 3, the share of hydroelectricity in the region falls substantially, as shown in table 4.1. Although there is some increase in the share of "other renewables," this increase does not come close to compensating for the projected fall in hydro. Maintaining the 2008 share of renewables in 2030 would require an increase in other renewable electricity of about 150 TWh (or about 7 percent of total generation in 2030).

The potential for expanded use of renewable electricity other than hydro has not been studied extensively at a regional level. It is clear that the physical potential of non-hydro renewable resources is far greater than that needed to maintain the overall share of renewables at today's levels until 2030. Serious consideration of a substantial increase in the

Table 4.1   Baseline Supply Scenario: Required Expansion by Latin America and the Caribbean to Maintain the 2008 Share of Renewables

Source	Generation (TWh)			Share of generation (%)		
	2008	2030	Change, 2008–30	2008	2030	Change, 2008–30
Hydropower	675	1,239	564	58.58	49.98	42.50
Other renewables	18	102	84	1.59	4.13	3.69
Total renewables	693	1,341	648	60.17	54.11	46.19
Total generation	1,153	2,480	1,326			
Additional generation required to maintain 2008 share of renewables			151			

Source: Authors' elaboration.

share of renewables by 2030 is also very relevant in light of the commitments made by some countries to reduce their greenhouse gas (GHG) emissions. The need to expand electricity supply will not stop in 2030. Therefore, for strategic planning, which includes mitigating climate change, it is also important to identify the renewable energy resources with the greatest potential to compensate for the inevitable decline in the region's hydropower share of new generation after 2030.

Wind

Short of a revolution in solar photovoltaic technology, wind would appear to be the renewable energy resource (aside from hydro) with the largest potential over the coming two decades. Development of this technology could fundamentally change the perception of the role that renewable energy could play in many countries' power expansion plans. A key question has been the extent to which a power system can absorb the inherent variability of the wind resource without excessive costs for backup. The cost per megawatt-hour (MWh) of the diverse "ancillary services" for the grid to provide backup to wind capacity will vary from one system to another and will tend to increase as the share of wind power in a given system increases. According to diverse studies published to date (for example, EERE/USDOE 2008 and NAS, NAE, and NRC 2009), a cost of US$5/MWh seems to be a conservative (high) value to cover all the increased costs of providing backup for wind power when it has reached a 20-percent share of capacity in thermally dominated power systems. Taking 20 percent of system-generating capacity as an upper bound with currently deployed grid technology would be equivalent to 9–14 percent of generation output, depending on the capacity factor. If one assumes this capacity ceiling and a regional output in 2030 of 2,500 TWh of generation, wind could supply up to 220–340 TWh.

Reaching the higher value of wind production by 2030 would be challenging, and probably would require a substantially greater level of electricity trade between countries in the region. Facilitating trade in wind would be an additional reason for the more extensive interconnections already recommended for the optimized development of hydropower in the region. Stronger intraregional interconnections are desirable to provide complementarity between different wind resources and with hydrologic resources. Greater interconnections would also help address the problem of intermittency. Although the output of individual wind turbines will vary considerably over the short term, the aggregation of a network of wind turbines over a larger area substantially reduces this variation, as has been shown in several European countries with large wind systems.

According to current estimates of wind resources in the region, which have not been extensively surveyed, the wind resource base in some countries may be too small to reach 20 percent of capacity. In other countries, such as Argentina and Mexico, there are abundant high-quality wind resources. The overall regional potential from high-quality resources is probably substantially larger than the 220–340 TWh "ceiling" set from the perspective of system integration. "High quality" means at least "Class 4" (7.0 meters per second or more) or even "Class 5" (7.5 meters per second)[3]—because higher-capacity factors can be obtained, which substantially reduces the cost gap with conventional fossil fuel–generation technologies.

The perception that wind energy is costly has reduced the interest in this technology by many regional power sector planners until recently. In the Brazilian energy auction in December 2009, in which 1,806 megawatts (MW) of wind were contracted, the costs for wind power were significantly less than expected. Indeed, the average cost of about US$83 per MWh[4] was slightly lower than the average price in some recent auctions in Brazil for fossil-fuel capacity.

With an estimated cost at about US$60 per MWh in 2008 (World Bank 2010), Mexico has some of the lowest costs of wind power generation in the world because of its high-quality wind resources. Unlike in Brazil, where wind must compete with relatively low-cost hydropower, in Mexico wind is quite attractive relative to the costs of electricity produced from natural gas, fuel oil, liquefied natural gas, and coal.

The variations in wind and hydroelectric output are largely independent of one another, which, together with the storage capabilities of hydropower dams, make wind and hydropower resources good complements. This complementarity makes it easier to incorporate wind energy in a power system with large amounts of hydropower, as demonstrated by the Nordic power exchange, Nord Pool Spot, and may increase the share of generation from wind that can be accommodated by the system.[5] It is important, therefore, to evaluate hydro and wind resources together—something that is not typically done. In the context of Latin America, where there is already a large share of hydropower and an even larger potential, wind power may be a logical choice for policy makers because of its complementarities with hydropower.

Biomass

Biomass, particularly residues from the sugar industry, currently provides a significant amount of electricity in Brazil and could make a further contribution in Brazil and elsewhere in the future. For countries using biomass from sugar, the potential depends on the amount of sugarcane harvested, the technology adopted by sugarcane mills, and the extent to which field trash (the tops and other sugarcane biomass that is left in the field after harvest) can be used. The current tendency in the sugar industry is to install power-generating equipment operating at 67 atmospheres (bar) pressure steam. More aggressive policies, which are becoming standard in Brazil and India, may raise the standard to above 80 atmospheres. At the same time, mechanized harvesting and prohibitions on burning field trash are opening the possibility of increasing the biomass fuel available for power generation. Table 4.2 provides an example of the amount of electricity that might be available to the grid in Brazil, depending on the technology, and the degree of use of field trash.

If 67-bar technology were used and one-fourth of plants used field trash for fuel, an increase of 55 TWh would be possible, compared with about 9 TWh sold to the grid in 2007. This is not far from the trend already underway in Brazil. If there were a jump to 82-bar steam technology and one-half of all plants used field trash for fuel, about 160 TWh of power might be available. If there were a technology breakthrough in gasification combined cycle plants, sales of electricity to the grid by sugarcane mills could increase even more.[6]

The values assume a sugarcane industry growing at 2.5 percent per year—roughly the historic rate of the previous decade—and reaching an output of about 1.2 billion tons of cane in 2030. This assumption implies that there is no large-scale expansion of sugarcane to supply ethanol or to

Table 4.2   Generation Options from Sugarcane Residues as Increase in TWh per Year Sold to the Grid, 2006–30

Trash use	67 bar	82 bar	BIGTCC[a]
25% of field trash	54.9	139.4	208.3
50% of field trash	64.9	161.9	251.8
75% of field trash	74.9	184.3	295.2

Source: Authors' elaboration.
Note: Data refer to the electricity that could be sold (or exported) to the grid and refer to the possible increase between 2006 and 2030. Sugarcane output is assumed to increase from 660 million tons to 1.2 billion tons. Data assume that all sugarcane output in 2030 goes to the specified option.
BIGTCC = biomass gasifier with combined cycle gas turbines, TWh = terawatt-hour.
a. The technology option BIGTCC is not yet technically proven.

satisfy global sugar demand (which would make more biomass available), and that the biomass residues are not diverted for the production of liquid biofuels from cellulose (which would make less biomass available for electricity production). Other sources of biomass (residues of the pulp and paper industry, rice, urban solid wastes), which have not been considered here, could also add to the expansion of electricity generation from biomass, though historically these sources have been relatively small. The important point is that if policies were put in place to promote biomass electricity, such as those in recent years in Brazil for sugarcane bagasse, there would likely be increasing supplies of biomass residues available for use in electricity generation.

Geothermal Power

Geothermal power is a proven technology and is currently commercially exploited in several countries in Latin America, both for direct heat and for power generation. However, the potential is very uncertain, especially for the region as a whole. Exploratory drilling has been limited, and the range of estimates is quite broad. For example, a review of geothermal power in Central America cites a range between 2 and 16 gigawatts (GW), with the most probable values being between 2.8 and 4.4 GW. If one extrapolates from the experience in the United States, where there has been a large amount of exploratory drilling, the potential of conventional geothermal resources in Latin America might be as much as 300 TWh per year. Of this, perhaps 125 TWh could be developed by 2030, compared with the 11 TWh currently generated.[7] The most important geothermal potential is concentrated along the tectonically active Pacific Rim from Mexico to Chile and in some Caribbean islands. Although capital-intensive, geothermal energy has high-capacity factors and can be quite competitive. It could be an important component of the power expansion in some of these countries if the commercial environment for developing projects can be improved.[8]

Solar Power

Solar electricity generation—mainly concentrated solar thermal power—will require breakthroughs to reduce costs if it is to contribute significantly in the next 20 years to grid-supplied electricity. Although small-scale photovoltaic technology (both off-grid applications and grid-connected building applications, including residential) will undoubtedly grow, it is unclear to what extent this will significantly add to the "other renewables" category of power generation in the region. Given the high cost of solar thermal power and the uncertainties with respect to large-scale (and grid-connected) solar photovoltaic development, no estimates of solar potential for power generation have been made in this report. In the near term, solar hot water applications could contribute significantly in a number of countries, as evaluated in both Brazil and Mexico, and, thus, replace large amounts of electricity and natural gas currently used for water heating.[9]

If one considers the potential of the non-hydro renewable resources discussed previously, it should be possible to maintain the current regional share of renewables (approximately 60 percent) even if it is not possible to increase the rate of expansion of hydroelectricity as indicated in chapter 3. Against a shortfall of about 150 TWh, the potential of new output of non-hydro resources might be in the following ranges:

- *Wind:* 220–340 TWh
- *Sugarcane:* 55–150 TWh[10]
- *Geothermal power:* 25–125 TWh
- *Solar power:* probably small for electricity generation (but potentially significant in hot water applications)

These estimates, especially the high estimates, assume concerted policies to promote the development of renewable resources. Nonetheless, the increase in the cost of generation relative to the projected costs outlined in chapter 3 could be quite small. Significantly increasing the regional share of renewables over the next two decades seems to be an economically feasible goal if governments are willing to put in place new policies promoting renewable energy resources, including the possibilities for greater interconnection and trade of electricity.

A Closer Look at Hydropower

With roughly 58 percent of the 60 percent of renewable energy in Latin America and the Caribbean coming from hydropower, the undeveloped hydropower potential in the region warrants a closer look. Table 4.3 provides estimates of the hydropower potential in the countries of the region and compares them with the hydropower capacity existing in 2007. The source of the information is the Latin American Energy Organization (Organización Latinoamericana de Energía, or OLADE), with adjustments made in the cases of Argentina and Brazil, where additional information is available.

Table 4.3   Potential Hydroelectricity Generation Compared with Existing Hydroelectricity Generation, 2007

Country or subregion	Existing hydroelectricity generation		Total potential hydroelectricity generation		Existing/potential (%)	
	Capacity (GW)	Output (TWh)	Capacity (GW)	Output (TWh)	Capacity	Output
Argentina[a]	9.94	31.06	40.40	130.00	25	24
Bolivia	0.49	2.32	1.38	4.81	35	48
Brazil[b]	76.94	374.38	251.49	1,213.00	31	31
Chile	5.37	22.80	25.16	26.56	21	86
Colombia	8.53	43.02	96.00	420.48	9	10
Costa Rica	1.41	6.77	6.41	28.08	22	24
Dominican Republic	0.47	1.68	2.01	8.80	23	19
El Salvador	0.47	1.74	2.17	9.48	22	18
Guatemala	0.78	3.01	4.10	15.21	19	20
Haiti	0.06	0.48	0.17	0.50	36	97
Honduras	0.50	2.30	5.00	21.90	10	11
Jamaica	0.02	0.17	0.02	0.11	90	162
Mexico	11.34	27.04	53.00	232.14	21	12
Nicaragua	0.10	0.31	1.77	7.74	6	4
Panama	0.85	3.87	3.28	14.38	26	27
Paraguay	8.13	53.71	12.52	54.82	65	98
Peru	3.23	20.03	58.94	385.12	5	5
Suriname	0.19	1.36	2.42	10.60	8	13
Trinidad and Tobago	0.00	0.00	0.00	0.00	0	0
Uruguay	1.54	8.07	1.82	7.95	85	102
Venezuela, RB	14.60	80.81	46.00	201.48	32	40
Total	**144.96**	**684.93**	**614.06**	**2,793.16**	**23**	**24**
Central America	4.10	18.00	22.70	96.80	18	19
Caribbean[c]	0.60	2.40	2.20	9.40	25	26

Source: OLADE 2008 (except Argentina and Brazil).
a. Devoto 2007.
b. BEN 2008.
c. Excluding Cuba.

The numbers presented in table 4.3 represent an overall optimistic view of hydropower potential. In addition, the definition of hydropower potential does not take into account the technical and economic feasibility of exploiting that potential. In some countries, the OLADE estimates may present more realistic or restrictive estimates of hydropower potential than others. Thus, there is great uncertainty about the potential size of the ultimate hydropower resource.

Nonetheless, it is clear that the remaining exploitable hydropower potential in the region will be significantly less than the value shown in table 4.3. Within the timeframe of 20–25 years, the feasible limit is even lower, although the major part of the remaining potential could be developed within this period. For illustrative purposes, this report assumes a minimum of 50 percent development of the potential in table 4.3 and a maximum of 70 percent for Latin America as a whole. Appropriate ranges for individual countries will be different and are not discussed here. Within this range, the quantity of remaining hydropower potential would be as follows:

- *50-percent development:* 712 TWh
- *60-percent development:* 991 TWh
- *70-percent development:* 1270 TWh

These values can be compared with hydroelectricity production in the region in 2007, which was 684 TWh. The potential does not cover the category of small hydropower for most countries. Although the definition of "small hydro" differs and the inventories of potential are less complete than for larger plants, the inclusion of small hydropower is unlikely to increase the overall hydropower potential by much more than about 5 percent.[11] Table 4.4 presents a different version of hydropower potential that conservatively assumes that only 50 percent of the remaining untapped potential in each country is viable.

It is important to distinguish between ultimate development and development that would be feasible in the next 20–25 years. Even the lowest-development scenario (50 percent of potential) would represent more than a doubling of the current capacity, which was built over a period of more than 50 years. One factor to consider is that more than half of the potential outside of Brazil is in countries that today use 10 percent or less of their resources (Colombia, Ecuador, Guyana, and Peru). Relying on the internal markets of these countries alone would be insufficient to absorb roughly half of their hydroelectricity potential over the next 20–25 years.

Case Studies of Renewable Energy Potential

The World Bank has recently completed detailed low-carbon development studies for two of the major countries in the Latin America and the Caribbean region. The renewable energy results of studies for Brazil and Mexico are presented here to provide an indication of the potential size of the untapped renewable energy resources. According to these and other detailed country studies, it is possible to infer that the renewable

Table 4.4   Potential Hydroelectricity Generation Compared with Existing Hydroelectricity Generation in 2007, Assuming 50 Percent of OLADE's Potential

Country or subregion	Existing hydroelectricity generation Capacity (GW)	Output (TWh)	Total potential hydroelectricity generation Capacity (GW)	Output (TWh)	Existing/potential (%) Capacity	Output
Argentina[a]	9.94	31.06	20.200	65.000	49	48
Bolivia	0.49	2.32	0.690	2.410	71	96
Brazil[b]	76.94	374.38	125.745	606.500	61	62
Chile	5.37	22.80	12.580	13.280	21	86
Colombia	8.53	43.02	48.000	210.240	18	20
Costa Rica	1.41	6.77	3.205	14.040	44	48
Cuba	0.04	0.12	0.325	0.650	12	18
Ecuador	2.06	9.04	11.875	48.385	17	19
El Salvador	0.47	1.74	1.085	4.740	43	37
Guatemala	0.78	3.01	2.050	7.605	38	40
Guyana	0.00	0.00	3.800	9.820	0	0
Haiti	0.06	0.48	0.085	0.250	36	97
Honduras	0.50	2.30	2.500	10.950	20	21
Jamaica	0.02	0.17	0.010	0.055	90	162
Mexico	11.34	27.04	26.500	116.070	43	23
Nicaragua	0.10	0.31	0.885	3.870	11	8
Panama	0.85	3.87	1.640	7.190	52	54
Paraguay	8.13	53.71	6.260	27.410	65	98
Peru	3.23	20.03	29.470	192.560	11	10
Dominican Republic	0.47	1.68	1.005	4.400	47	38
Surinam	0.19	1.36	1.210	5.300	16	26
Trinidad and Tobago	0.00	0.00	0.000	0.000	0	0
Uruguay	1.54	8.07	0.910	3.975	85	102
Venezuela, RB	14.60	80.81	23.000	100.740	63	80
Total	**144.96**	**684.93**	**307.03**	**1,396.58**	**46**	**48**
Central America	4.10	18.00	11.350	48.400	36	37
Caribbean[c]	0.60	2.40	1.100	4.700	55	51

Source: Authors based on data from OLADE 2008 (except Argentina and Brazil).
a. Devoto 2007.
b. BEN 2008.
c. Excluding Cuba.

energy potential in other countries of the region is likely to be significant and that the current planning estimates understate the potential.

Brazil. For the Brazil low-carbon study (World Bank 2010), the renewable energy-based mitigation options that were analyzed include wind, sugarcane bagasse, and solar energy for water-heating systems. For wind generation, the national energy plan (known as PNE 2030) foresees a tenfold increase in capacity—reaching 4.7 GW and supplying about 1.5 percent of Brazil's electricity in 2030. This estimate assumes that a number of barriers can be overcome, including high upfront investment costs and the short-term difficulty of procuring equipment. Both of these barriers relate in part to the current structure of public auctions for renewable energy in Brazil that includes a 70-percent domestic content rule (under PROINFA [Programa de Incentivo às Fontes Alternativas de Energía Eléctrica]), which has delayed the scheduling of wind power projects and raised costs given the small number of local manufacturers of turbines and components. If one assumes that these barriers can be overcome, and according to projections made by the Brazilian Wind Energy Association, the low-carbon study estimates that the expansion of installed wind capacity could be as high as 15 GW by 2030. With regard to sugarcane bagasse cogeneration, the study suggests that about 40 GW of installed capacity would be available by 2030 that could provide electricity to the grid—compared with 6.8 GW in the reference scenario—corresponding to about 100 TWh of electricity being generated in 2030.

PNE 2030 already assumes that hydropower will represent close to 70 percent of power generation in 2030, implying an unprecedented increase in the production of hydropower and virtually the full exploitation of Brazil's hydroelectricity potential. The low-carbon study explores the option of increasing hydropower production even further. By interconnection of the electricity systems of Brazil and República Bolivariana de Venezuela, whereby the existing and planned hydropower plants located in hydrologically complementary regions in the Amazon Basin would be linked, there could be an exchange of 21.7 TWh of power between the two countries and the displacement of thermal plants that are currently used for providing power during the low periods in both countries (so-called "valley filling").

In terms of the financing needs for implementing a low-carbon strategy in the electricity subsector, investment requirements for 2010–30 amount to US$66.0 billion: US$52.3 billion of investment would be required for sugarcane bagasse cogeneration, US$12.9 billion for wind-based generation, and about US$0.45 billion for the transmission line connecting Brazil and República Bolivariana de Venezuela.

The study outlines several policy and regulatory measures that would help increase the participation of renewable energy sources in Brazil's electricity mix. For hydropower, the simplification of the environmental licensing process is considered critical. For bagasse and wind-based genera-tion, an important barrier is the cost of interconnection with the some-times distant or capacity-constrained subtransmission grid, with a key question being the responsibility for financing the grid connection. The study recommends the ambitious development of a smart-grid program as a way to optimize the contribution of distributed wind- and bagasse-based electricity generation potential. For expansion of the use of solar energy, it is recommended that the industrialized products tax on solar energy products—such as for solar thermal hot-water collectors and solar photovoltaic panels—be substantially reduced.

Mexico. The Mexico low-carbon study (Johnson and others 2010) evalu-ates 40 low-cost interventions across key emissions sectors in Mexico, developing a low-carbon scenario through 2030 and assuming no major improvements in technology or reductions in technology costs. Among the renewable energy technologies evaluated for Mexico were those offering baseload (geothermal), intermittent (wind), and peak generation (biomass, most small hydropower, and cogeneration).

According to the baseline scenario, even at a net cost of carbon diox-ide (CO_2) of as little as US$10 per ton, additional low-carbon energy technologies, such as small hydro resources, wind, biomass, geothermal, and cogeneration, could replace much of Mexico's fossil-fuel generation. Under a low-carbon scenario, the contribution of renewable energy increases substantially—from 1.4 percent to 6 percent for wind— (primarily by developing wind farms in Oaxaca), with additional wind capacity by 2030 of 10.8 GW; from 2 percent to 11 percent for geo-thermal (7.5 GW); and from 14 percent to 16 percent for hydropower (2.7 GW). Other renewable and nonconventional energy sources in the low-carbon scenario for Mexico include cogeneration from Pemex facili-ties (3.7 GW), industry (6.8 GW), sugar mills (2 GW), and other biomass (5 GW) that would provide 13 percent of new power capacity, and through fuelwood co-firing in existing plants (2.1 GW). The projections are based on calculations that compare the net costs of each renewable energy technology with the costs of the displaced fossil-fuel capacity (mainly natural gas and coal).

Relative to the baseline, an estimated US$10 billion of net invest-ment would be required to implement the low-carbon scenario in the power sector. The corresponding new investment required in wind-based generation over the 2010–30 timeframe would amount to US$5.5 billion, whereas US$1.1 billion would be required for biogas electricity,

US$2.6 for small hydropower, US$11.8 for geothermal, US$3.0 billion for cogeneration in Pemex, and US$3.7 billion for cogeneration in industry.

The Mexico low-carbon analysis assumes that the majority of investment (under independent power producers) in the power sector would be undertaken by the private sector, with the public sector (through the state monopoly Comisión Federal de Electricidad [CFE] and regulatory agencies) continuing to play a major role. To reach the increased contribution of renewable energy in Mexico's electricity mix, the low-carbon analysis identifies a number of policy changes that would need to be implemented, including removing the regulatory restrictions that currently exist for small-scale and renewable energy technologies. Other key obstacles to renewable energy development in Mexico include the absence of externalities for new fossil fuel–based power generation and the lack of capacity payments for intermittent energy sources such as wind.

Regional Trade and Cross-Border Integration

Regional electricity trade could play an important role in allowing Latin America and the Caribbean to meet its growing electricity needs in a cost-effective way and could also help to foster the expanded use of regional energy resources. By linking countries and regions, interconnections allow the optimization of electricity supplies, which can improve efficiency and may reduce the need for domestic investment in high-cost generation capacity and backup. As discussed in the previous section, the development of large-scale hydroelectric, wind, and other energy resources could benefit through increased trade by expanding the market into which the projects would sell. This section discusses the benefits of trade as well as the barriers that inhibit greater regional market integration. The section also summarizes the basic characteristics of the regional electricity market operating in Latin American countries today and the new and planned interconnections that could allow greater electricity trade. The section also presents the results of an electricity trade simulation exercise for Central America that provides preliminary estimates of the benefits of regional trade.

As shown in chapter 2, there are already a number of electricity interconnections in Latin America and the Caribbean and new ones are being developed. Nonetheless, it has also been seen that electricity trade plays a very small role in meeting supply needs in the region. For trade to increase, it is important to overcome countries' fears that supply contracts will be broken and energy supplies interrupted. Although there are examples in the region—for both electricity and natural gas—where trade has been interrupted, such cases are the exception rather than the rule, and it is important to move forward with clear principles and rules to reduce the risks associated with regional trade.[12]

Benefits of Cross-Border Integration

In principle, regional interconnection and electricity trade between countries is an attractive approach for expanding the supply of electricity:

- Trade can enhance the reliability and security of the local network by linking it with a larger grid and a greater number of generation sources, thus increasing the diversity of the generation system.
- Trade may reduce generation costs because of economies of scale associated with power generation from larger facilities. Optimizing capital requirements for the electricity sector can free up capital resources for other investments and improve the domestic fiscal situation.
- Interconnection (and the ability to acquire power through trade) allows individual countries to have lower "reserve" requirements, which reduces the need (and cost) of investing in reserve power capacity.
- Trade may allow more competition in open markets as it increases the availability of electricity from different sources at varying costs. In addition, interconnections between markets may allow for some convergence of electricity prices, because the connected areas can function as a single market. Interconnection may lead to an important reduction in variable costs because countries do not need to import fossil fuels.
- In the case of seasonal renewable resources such as hydropower, interconnection allows the linking of basins with different hydrology. This linking increases the firm energy that can be supplied by the same set of dams. This balancing of variable renewable resources also applies to wind and even biomass energy.

Despite the potential benefits of interconnections, electricity trade, and cross-border electricity projects (those that rely on multicountry markets), there have been significant political and regulatory barriers in Latin America and the Caribbean that have hindered trade. When a project is planned and begun across borders, there are likely to be different technology standards, regulatory regimes, pricing policies, environmental concerns, and legal frameworks. More significant, there can be different views about investment costs and the way they are shared. Nevertheless, such issues can be resolved if there is a clear economic and commercial motivation behind the project that benefits all countries. Other issues that can affect project development are market changes; the emergence of new sources of fuel or electricity; and demand shocks, such as a financial crisis, that may dramatically alter the conditions for trade. Ultimately, the greatest uncertainties tend to be connected with political decision making, and these may be particularly difficult to predict or address.

In the context of climate change, regional interconnection could expand the share of renewable energy in the electricity supply mix, and this benefit of regional trade deserves more attention than it traditionally has received. Brazil's experience in expanding its hydropower resources is instructive. Brazil, which has 40 percent of the estimated hydro resources in the region, is responsible for 54 percent of the region's hydropower generation. Although about 84 percent of Brazil's electricity is generated from hydropower,[13] that share is only 27 percent in the rest of the region. Under the ICEPAC (Illustrative Country Expansion Plans Adjusted & Constrained) scenario, this differential would be maintained. The large disparity in the development of hydropower does not appear to be a result of Brazil's superior hydro resources. (Indeed, the remaining resources outside Brazil are probably substantially larger.) A more likely hypothesis is that by integrating at a continental scale (almost unique in the world), Brazil has been in a better position to exploit hydropower resources than the relatively small and fragmented markets in the rest of Latin America.

In Latin America, there are three primary clusters in which electricity trade currently occurs: (a) Central America and Mexico; (b) Colombia, Ecuador, and República Bolivariana de Venezuela; and (c) Argentina, Brazil, Paraguay, and Uruguay.

Mexico and Central America Interconnection

In 1996, the Central American countries agreed to the creation of the Regional Electricity Market (Mercado Eléctrico Regional, or MER) through the Framework Treaty for the Central American Electricity Market. To support the MER, the treaty also created a regional regulatory commission (Comisión Regional de Interconexión Electrica, or CRIE), a regional system operator (Ente Operador Regional), and a company owning the grid (Empresa Propietaria de la Red, or EPR). Map 4.1 illustrates the existing Central American regional transmission grid (Red de Transmisión Regional, or RTR) as it was defined during the transition period up to the commissioning of SIEPAC (Sistema de Interconexión

Map 4.1   Regional Transmission Grid in Central America, 2006

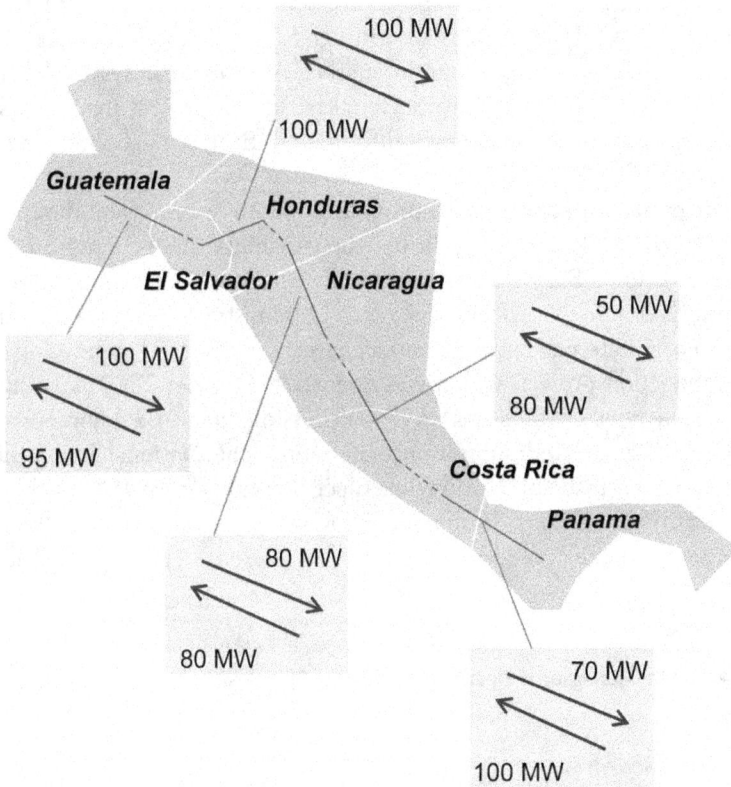

Source: Authors based on information from Comisión Regional de Interconexión Eléctrica (CRIE).

Eléctrica de los Países de America Central). The system consists of individual 220 kilovolt (kV) interconnection links connecting the power systems of neighboring countries and is used to provide short-term international power exchanges.

To reinforce the regional interconnection, the main transmission companies of Central America (as well as ENDESA [Empresa Nacional de Electricidad S.A.] in Spain, ISA [Interconexión Eléctrica S.A. E.S.P.] in Colombia, and CFE in Mexico) are participating in the regional transmission company, EPR, through the SIEPAC project, which is in charge of the reinforcement of the RTR. The system consists of a 230 kV interconnection system with 300 MW of transmission capacity between the countries. The project interconnects the six countries in Central America into a single system. The countries agreed to strengthen the interconnectivity of the region and create a standard regulatory framework for the power sector. Map 4.2 illustrates the SIEPAC project.

In 2007, the interregional trade through SIEPAC represented less than 2 percent of the total supply to the market. Although the system has not had a significant effect on intraregional trade, the project itself is a significant achievement for the region from a political, regulatory, and technical point of view.

The connection with Mexico provides a link to Mexico (and the rest of North America) that can help buffer the variability of electricity production in individual SIEPAC countries.[14] The Mexican Secretariat of Energy (Secretaría de Energía , or SENER) and the Ministry of Mines and Energy (MME) of Guatemala entered into an agreement in 2003 to develop the interconnection project, which was opened in October 2009. The initial connection consists of a 400 kV transmission line, 103 kilometers in length with associated substation expansions. The National Institute of Electrification (Instituto Nacional de Electrificación, or INDE) has already contracted with CFE to purchase 120 MW, and it is expected that the remaining capacity of the line will be traded in the Guatemalan

Map 4.2   SIEPAC Regional Electricity Exchange

Source: Authors based on information from SIEPAC.

Opportunity Market. By connecting to Mexico, SIEPAC establishes a link northward to a major electricity producer.

Colombia-Panama Interconnection

The Colombia-Panama interconnection connects Central America with its southern neighbors, with the potential to tap low-cost hydroelectricity and other energy resources in the future. The Colombian and Panamanian authorities agreed to construct a 614-kilometer power transmission line at 250–400 kV (high voltage direct current) and with the capacity of 300 MW and possible expansion to 600 MW. Investment costs are estimated at US$210 million, including the required expansion of the substations. The project is under construction, and commissioning is expected to take place in 2013. The execution of this project opens the possibility for the physical connection of the electricity markets between the Andean countries and Central America and Mexico.

South America Integration Potential

The electrical interconnections in South America are concentrated in two clusters. The northern cluster includes Colombia, Ecuador, and República Bolivariana de Venezuela, and the southern cluster includes Argentina, Brazil, Paraguay, and Uruguay. Currently, a capacity of more than 20 GW is available for regional trade in the southern cluster. More than half of that capacity is attributable to Itaipu (12.6 GW). Itaipu, Rincon-Garabí, and Yacyretá interconnections concentrate more than 80 percent of the total capacity available for interregional trade in the southern cluster. The three facilities link Argentina, Brazil, and Paraguay.

Table 4.5 summarizes the basic characteristics of the binational power plants in South America, and table 4.6 shows the international power

Table 4.5 Binational Power Plants in South America

Countries	Name	River	Installed capacity (MW)	Status
Bolivia–Paraguay	Itaipu	Parana	12,600	Operating
			1,400	Additional capacity
Argentina–Uruguay	Salto Grande	Uruguay	1,890	Operating
Argentina–Paraguay	Yacyretá	Parana	2,100	Operating
			3,100	Additional capacity
Argentina–Brazil	Garabi	Uruguay	1,500	Evaluation
Argentina–Paraguay	Corpus	Parana	3,400	Evaluation

Source: Comisión de Integración Energética Regional 2008.
Note: The additional 1,400 MW of Itaipu have been operating. Of the additional 3,100 MW of Yacyretá, 2,000 MW are unused existing capacity and 1,100 MW are additional capacity under construction.

Table 4.6 International Power Interchanges in South America, 2007 GWh

Importer	Exporter							Total Imports
	Argentina	Brazil	Colombia	Ecuador	Paraguay	Uruguay	Venezuela, RB	
Argentina		1,999			7,479	971		10,449
Brazil	5				37,936	34	537	38,512
Chile	1,628							1,628
Colombia				38			6	44
Ecuador			877					877
Uruguay	574	215						789
Total exports	2,207	2,214	877	38	45,415	1,005	543	52,299

Source: Comisión de Integración Energética Regional.
Note: GWh = gigawatt-hour.

exchanges among interconnected countries in 2007. International power interchanges reached 52 TWh in 2007, of which 45 TWh (about 90 percent) were associated with Argentina and Brazil's power purchases from the Yacyretá and Itaipu binational power plants.

Latin America has a diversity of interconnections for the regional trade of electricity as shown in table 4.5. Electricity trade requires not only physical investments in transmission systems, but also regulatory processes governed by regional, as opposed to national, rules. The experience of success and failure with the interconnection agreements is quite rich and diverse and is useful for planning and promoting future projects in the region. Although the possibilities for new interconnections are ample and have the potential to boost interregional trade, there are obstacles to execution. Most countries in the region have been reluctant to give up national regulatory processes for the pursuit of potential regional gains. Successful trade expansion and collaboration among countries depend critically on the real and perceived economic and political costs and benefits.

Risks and Potential Barriers to Cross-Border Integration

Economic feasibility is a necessary condition for implementation of cross-border electricity trade projects, but it is not sufficient. A recent study (Soreide, Benitez, and Haladner 2009) shows that the economic rationale for a proposed interregional electricity supply project is a critical condition for success. In addition, there are other conditions that must be met to overcome nonmarket obstacles. Factors that may help with the implementation of interconnection projects are an appropriate regulatory framework and, ultimately, political will.

Cross-border, long-term investment involves a high degree of uncertainty for both governments and investors. In the case of Latin America, cross-border electricity supply agreements have always been subject to changes in incentives and conditions for trade, such as shifts in political regimes, in political priorities, and in the economics of the projects themselves. Such changes apply to both supply and demand in long-term, interregional trade agreements.

On the diplomatic front, frictions between countries often affect trade agreements including cross-border electricity contracts. This was the case with natural gas contracts in South America, where diplomatic disagreements made some countries more vulnerable. Furthermore, because of such risks and the associated implications to energy security, countries opt for developing further domestic energy supplies, which are not necessarily the most efficient option (even accounting for geopolitical risk). To ease such risks, deeper and broader diplomatic relations would be required, though some of the risk might persist. Box 4.1 presents a summary of the potential obstacles facing cross-border electricity supply projects.

Box 4.1

Potential Obstacles to Cross-Border Electricity Exchanges

Cross-border electricity supply faces the following potential obstacles:

- **Changing motivations:** Strategic factors influence choice of buyer or seller.
- **Cost-recovery:** The debt burden from the project becomes politically difficult.
- **Financial crisis:** Critical changes occur in the macroeconomic environment.
- **Free-rider problems:** One party involved avoids its share of investment costs.
- **Inappropriate regulatory framework:** There is a weak regional or bilateral context for collaboration.
- **Incomplete contracts:** Different views on contingencies are not covered by a contract.
- **Incentive problems:** Those responsible for risk are not the same as those who bear the burden.
- **Institutional framework:** Red tape causes obstacles to monitoring efforts, maintenance, and trade.
- **Market changes:** Alternative sources of energy are available, or there is higher demand in the exporting country.
- **Political instability:** A shift in the political regime leads to changes in the commitment to the deal.
- **Private agendas in politics:** Private agendas influence the design and function of the project.
- **Sunk cost:** Terms of the deal are altered after investments have been made.
- **Vested interests:** Those who benefit from the status quo oppose cross-border collaboration.
- **Weak property rights and contract enforcement:** Deals are not upheld as intended, and contracts are not enforced.

Source: Soreide, Benitez, and Haladner 2009.

Interregional trade would allow countries to diversify their sources of power generation in addition to the gains from trade associated with specialization from the most efficient producers. Moreover, a regional approach could reduce the emission of GHGs as the share of renewables, and low GHGs would increase through greater integration. Current interconnections and those being actively studied (as shown in map 4.1) would tap only a fraction of the complementarity of hydro resources inherent in the fact that South America straddles the Equator. An example of an "interhemispheric" connection was evaluated in the Brazil low-carbon study (World Bank 2010). In the case study example, the Simon Bolivar hydro plant (formerly Guri) on the Caroní River would be linked to the proposed Belo Monte plant on the Xingu River, both with a capacity over 10 GW. The two rivers' seasonal hydrologies are almost a mirror image of each other, and the interconnection would effectively link the two national grids. A preliminary estimate is that almost 22 TWh could be exchanged with gains for both sides, including substantial financial benefits and reductions in GHG emissions.

A key determinant of the success of an interregional approach to the cross-border supply of energy is the regulatory framework, which among other things involves technical factors (such as types of voltage and frequency in place); price structures; market operations; and openness to private sector investment, contract enforcement, and monitoring and environmental aspects. In fact, investors look for compatible regulatory frameworks in the pursuit of regional projects. Price differences and variations in the rules across countries make the harmonization of interconnection protocols in the power sector difficult.

Evaluating the Impact of Interconnections: An Example from Central America

What are the likely consequences of increased electricity trade? This section presents an estimate of the effect of interconnecting the electricity grids of Central America based on the results of the modeling exercise carried out in chapter 3. This exercise evaluates the supply and demand investment requirements of the unified subregion of Costa Rica, El Salvador, Guatemala, Honduras, Nicaragua, and Panama. In the scenario with no trade, each country minimizes the costs of meeting its electricity requirements by itself. In the trade scenario, a cost-minimization exercise is conducted for the Central America region as a whole, essentially assuming that the subregion behaves as one market. A key assumption of the trade scenario is that the SIEPAC interconnection transmission system is complete. Other assumptions within the model, including fuel prices, electricity demand, technology costs, and resource availability, are the same for both scenarios.

Figure 4.1 shows the results of the trade scenario for Central America. A principle factor that changes the results for the trade scenario is the economies of scale of integrating the subregion. This factor leads not only to a mix of technologies that is different from the generation mix without trade, but also to lower average generation costs as a result of trade. The modeling exercise indicates that Central America would have a higher percentage of hydropower under a trade scenario and a lower percentage of fuel oil plants. The increased hydropower comes mainly from the hydropower resources available in Costa Rica, El Salvador, and Nicaragua. If Central America were integrated with both its southern and its northern neighbors (something not carried out in the exercise), there would be even more room for trade and presumably lower supply costs. The subregional market scenario would also allow for a decrease in the redundant, recurrent expenses that countries face by maintaining higher reserve capacity, and would lower fuel imports (and the variability that has accompanied fossil-fuel prices).

Another implication of the results is that the subregion would become less carbon-intensive over the forecast period. Figure 4.2 illustrates the

Figure 4.1   The Impact of Trade in Central America

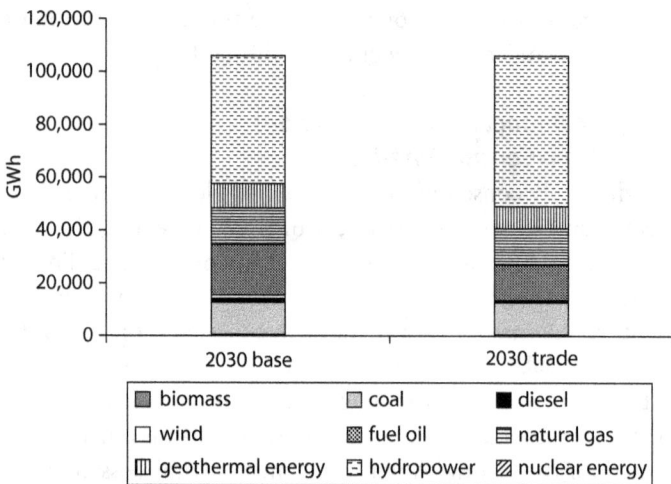

Source: Authors' elaboration based on optimization model.
Note: GWh = gigawatt-hour.

Figure 4.2 CO_2 Emissions in the Base Case versus the Trade Case

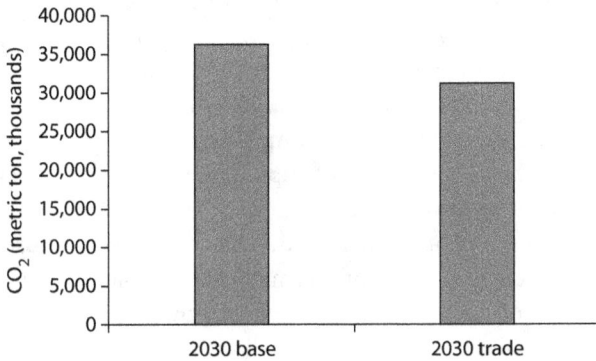

Source: Authors' elaboration based on optimization model.

levels of CO_2 emissions for both the base case and the trade case. Emissions from the trade scenario are lower largely as a result of a higher share of hydropower in the trade scenario's generation mix.

Energy Efficiency

Of all the options for meeting future electricity supply, energy-efficiency measures are almost always the least expensive (from a financial and economic perspective). And despite the growing awareness of, and focus on, energy-efficiency benefits, a gap remains between energy-efficiency potential and implementation. This section presents a brief review of some of the supply-side and demand-side energy-efficiency options available in Latin America and the Caribbean, including estimates of how much energy efficiency might contribute to the future electricity supply needs of the region.

Supply-Side Efficiency: Options and Potential

Supply-side energy efficiency in the electricity market essentially encompasses all the measures that can help conserve or save energy in the production, transport, and delivery of electricity.[15] Over the past few decades, countries in the region have had diverse experiences implementing reforms and programs aimed at improving supply-side energy efficiency in the electricity sector. For instance, during the 1990s Argentina was able to achieve efficiency improvements in the production, transmission, and distribution of electricity as a result of the electricity reform process, which contributed to the country's power and gas sector being among the most competitive in the region. A range of supply-side efficiency measures is given in table 4.7.

Investments in power transmission, including those with private sector participation in construction and maintenance of lines, offer a large potential for energy savings. For example, power transmission improvements through the development of superconducting power transmission cables are estimated to help achieve energy savings of up to 40 percent for high load connections, whereas the most important losses in such cable systems occur as a result of thermal insulation and cooling-machine inefficiencies. Especially in countries where power generation is located far from the demand centers, even a small percentage reduction in power transmission losses over thousands of kilometers can save considerable amounts of energy and can provide a more cost-effective alternative than investments in additional generation.

In geographically large countries, improving the efficiency of transformers and reducing the instances of overloading has proven important for enhancing the quality of power supply, as has the use of ultra high voltage, which can deliver large quantities of power with very little loss. Transmission systems intended for long distances—in countries such as Brazil—generally need to incorporate both advanced hardware and software technologies to improve power transfer and increase the use of the facilities already in place, thus avoiding additional transmission capacity investments and environmental impacts.

Table 4.7   Measures for Increasing Supply-Side Energy Efficiency

	Measure
Power generation	Plant rehabilitation and refurbishment; improved operation and maintenance practices and better resource use (higher plant load factors and availability) in existing generation facilities; new thermal power plants (combined cycle, supercritical boilers, and IGCC); fuel switching; co-generation or combined heating, cooling, and power (CHP)[a]
Power transmission and distribution	Efficient and low-loss transformers; high-voltage lines; improved insulation of conductors; use of capacitors; improved metering systems and instrumentation; substation rehabilitation; Smart Grids or system optimization.

Source: World Bank 2009a.
Note: IGCC = integrated gasification combined cycle.
CHP uses more than 80 percent of the useful energy in fuel, compared to consuming 35–50 percent of useful energy in the case of power generation alone.
a. Expert Group on Energy Efficiency 2007.

The Latin America and the Caribbean region's countries differ markedly in terms of both the absolute levels and the trends in supply-side energy efficiency and, particularly, in the electricity losses and associated costs resulting from inefficiencies in power transmission and distribution. As shown in figure 4.3, electricity distribution losses in Latin America range from as low as 6–7 percent in Chile to about 15–20 percent in most other countries of the subregion. Also, for the region's biggest electricity consumer, Brazil, distributional losses in the past seven or eight years have fluctuated between 13 and 15 percent. Paraguay has been an outlier among the countries of the region in terms of the absolute level of distribution losses, where the figures have shown a steep increase over the past several years from about 22 percent in 2001 to above 30 percent in 2005.

The average energy efficiency of electricity distribution in most of Central America and Mexico has been in a range similar to that of the rest of the region. Electricity losses have been relatively stable over the past five years, ranging from about 10 percent in Costa Rica to about 17 percent in Panama (figure 4.4). Exceptions from this general trend are Nicaragua, which has had high though somewhat declining levels of distribution losses, and Honduras, which has experienced levels around 30 percent. The electricity sector in Mexico has been characterized by overall distribution losses that compare to most other countries in the region. At the same time, however, its level of technical losses—at about 10 percent—is relatively high, compared with not only countries of similar average per capita incomes, such as Brazil and Chile, but also El Salvador and Paraguay. Before being taken over by CFE, the electricity utility LFC (Luz y Fuerza del Centro), serving the greater Mexico City area, had distribution losses several times higher than the national average and among the worst in the world (Komives and others 2009). The company was liquidated in fall 2009 largely as a result of the high losses and large subsidies required from the federal government to keep it afloat.

In the Caribbean subregion, the level of absolute distribution losses is comparable to those of the Southern Cone and Central America. The exception is the Dominican Republic, where losses are about 10 percent higher than in the rest of the region and have been increasing in recent years for which data are available. Similarly, in Jamaica, where the electricity sector represents about 23 percent of overall energy consumption, losses have been on the rise between 2002–05, although there are indications that they may now be on the decline.[16]

Figure 4.3 Total Distributional Electricity Losses in Latin America and the Caribbean

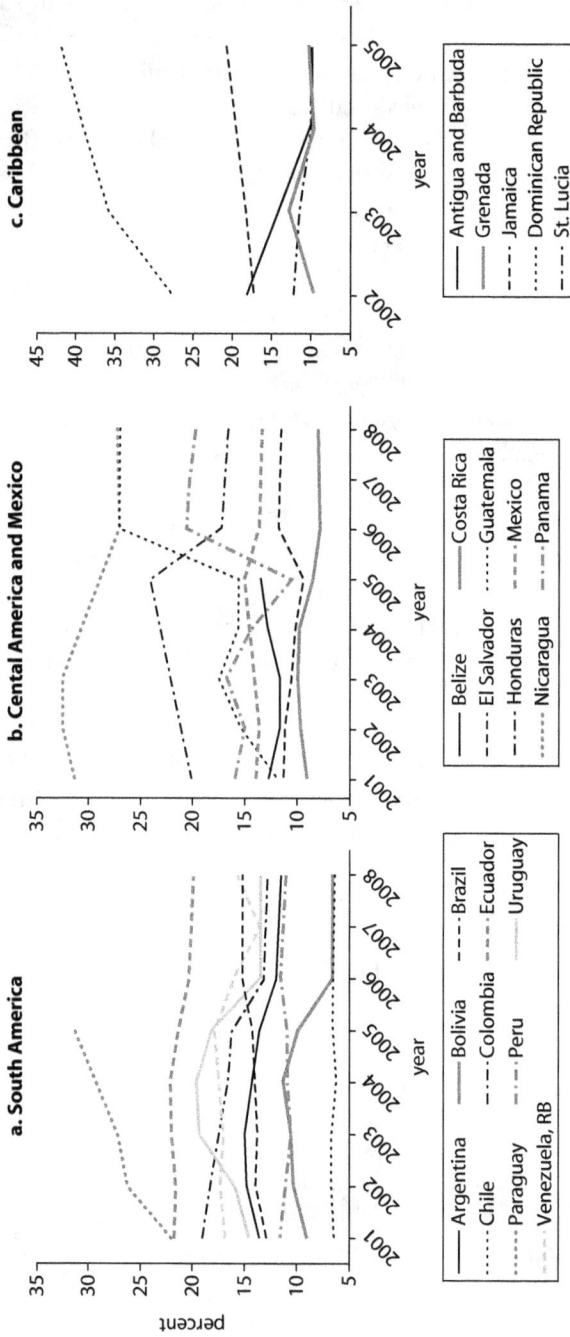

Source: Authors' calculations based on OLADE 2009 (for 2006–08); Benchmarking Data of the Electricity Distribution Sector in the Latin America and the Caribbean region, 1995–2005 database, World Bank.

Figure 4.4   Annual Electricity Sales and Distributional Electricity Losses, 2005

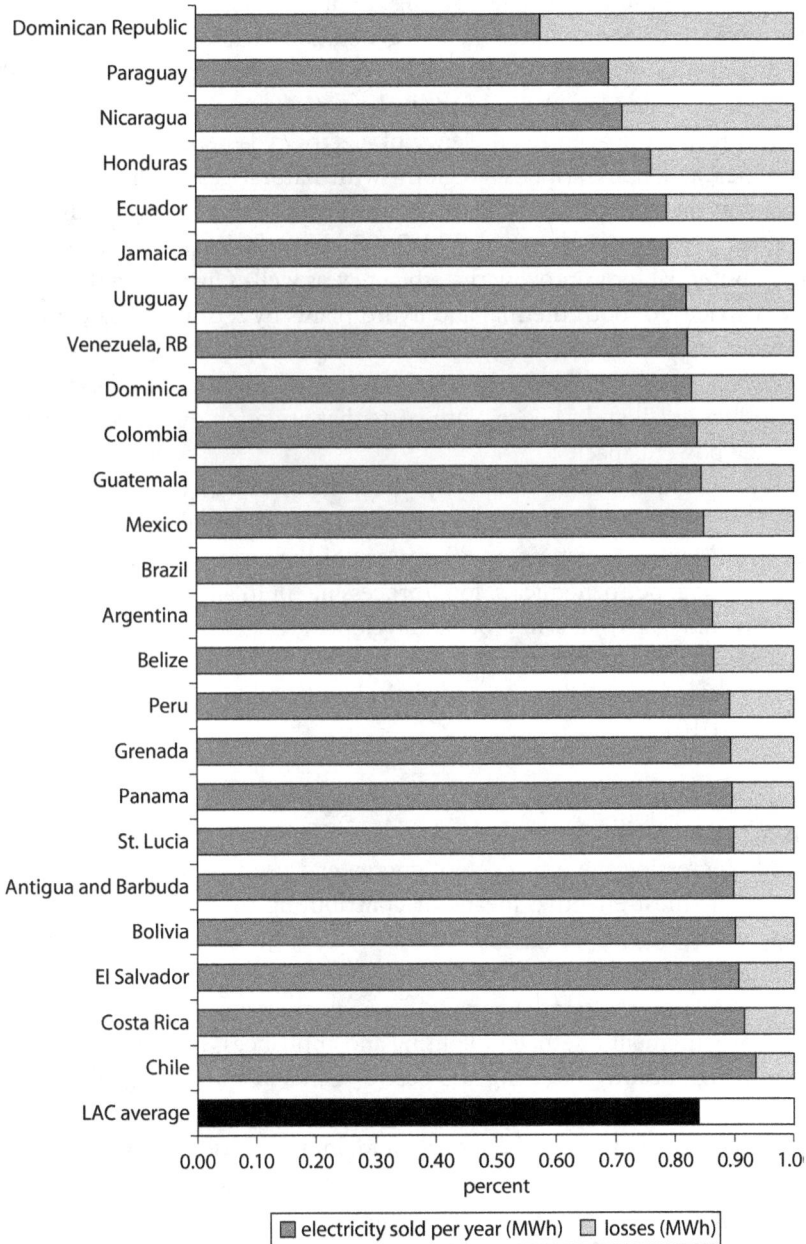

Source: Authors' calculations based on Benchmarking Data of the Electricity Distribution Sector in the Latin America and the Caribbean region 1995–2005 database, World Bank.
Note: LAC = Latin America and the Caribbean.

In addition to transmission and distribution improvements, there can also be large energy-efficiency savings by improving the technology and the operations and maintenance of generation facilities. Because of the higher average efficiency of natural gas plants (at 40 percent, on average) compared with coal- and oil-based plants (34 percent and 37 percent, respectively), the average efficiency of electricity generation in the Latin America and the Caribbean region—with natural gas as the dominant fuel for thermal plants in most countries—is already higher than in regions relying on coal and oil (IEA 2008). Nonetheless, there is typically large potential for improving the efficiency, as well as increasing the effective capacity, of both thermal and hydro plants by retrofitting key power production or auxiliary equipment. Large "repowering" investment opportunities have been identified in major countries in the region, including Brazil and Mexico, with costs that are often a fraction of those of new power capacity.

Cogeneration—using the waste heat from electricity or heat-only applications to generate electricity—has great potential in many countries. In Mexico, an estimated 80 percent of the cogeneration in industry (oil refining, petrochemicals, food processing, pulp and paper, sugar, and textile and glass) has not been developed. Cogeneration in Pemex's facilities alone is estimated to be more than 3,600 MW, or more than 6 percent of Mexico's total installed electricity capacity (Johnson and others 2010).

Demand-Side Efficiency: Options and Potential

In addition to being the least-cost way of meeting future electricity needs, increasing the efficiency of energy end use has been identified as the single-most-effective means of contributing to the goals of energy supply security, improved affordability of energy services, and environmental sustainability (IEA 2008). In countries with effective and conducive regulatory systems, energy-efficiency programs (such as through bulk procurement of efficient lighting and appliances) have proven effective in significantly reducing the cost of demand-side energy-efficiency measures. Efficiency programs run by electric power utilities have been introduced in a number of countries and, in many cases, have resulted in significant reductions in electricity bills and the deferring of investments in new generation capacity for power utilities. There is a range of ways to promote demand-side energy efficiency. One of the key principles working in favor of demand-side energy efficiency is that the cost of conserved energy is lower than the cost of new power-generating capacity.

As summarized in table 4.8, numerous instruments are available for improving demand-side energy efficiency. These instruments include national energy-efficiency resource standards (utility energy-saving targets); energy codes for new buildings; appliance standards; national and state-level energy-efficiency tax incentives; programs to promote comprehensive energy retrofits to existing buildings; energy-efficiency labeling and disclosure programs; and support for the dissemination of solar water heating and efficient appliances. As outlined previously, these measures generally fall into two categories: (a) those aimed at changing the

Table 4.8   Demand-Side Energy Efficiency Instruments

	Instrument
a) Load management	Demand charges; direct load control; demand response programs; tariff incentives and penalties (for example, power factor penalties, time-of-use rates, and real-time pricing)
b) End-use energy efficiency	
Industrial	Energy audits and performance measurements; energy efficiency financing, services provided by ESCOs; combined heat and power (co-generation); fuel switching; waste heat recovery; efficiency improvements of industrial motors and drive systems; equipment regulations and standards; monitoring and verification of system-wide energy flows
Buildings	Integrated building design; building codes; building retrofits; envelope measures (insulation and windows); efficiency standards for lighting; use of passive lighting; efficient pumping; efficient space, water heating, and cooling; application of solar water heaters and passive space heating; timers and temperature controls on electric hot water cylinders; reduced standby losses in appliances and equipment; energy management systems
Residential	Building codes; appliance standards; labeling; consumer education; improved cooking stoves; improved district heating (for example, through boiler rehabilitation, pre-insulated piping, compensators, pumps, and heat exchangers)
Public	Efficient street lighting; efficient water pumping and sewage removal systems; combined heat and power; "watergy" (that is, energy and water efficiency in water supply and wastewater treatment); internal information campaigns to promote energy efficiency and best practices in the operations of public agencies
Agriculture	Efficient irrigation pumping and drip irrigation; efficient agricultural equipment

Sources: World Bank 2009a, UNIDO and REEEP 2010.
Note: ESCO = energy service company.

load pattern and encouraging less demand at peak times and peak rates, and (b) those implemented to reduce demand through more efficient processes, buildings, or equipment.

Load management is used by utilities to relieve constraints on distribution and transmission networks and, in the long term, can defer the need for new power capacity. By redistributing the load, such as by moving consumers from peak to off-peak hours, utilities can lower the costs of producing electricity by deferring the use of high-cost peak load. Because peak-load plants in some countries rely on diesel or fuel oil, load management can also lower the carbon-intensity of generation. However, unlike energy efficiency, load management does not directly result in less electricity being generated. Load-management programs, although generally easier to implement than other demand-side management (DSM) strategies not under the direct management of utility companies, are largely short-term responses that have direct financial benefits to the utilities. But load management is a small part of the total demand-side potential, and a combination of load-management programs with end-use energy-efficiency programs can raise the effectiveness of both approaches and lead to greater demand reductions.

Increasingly used as part of a comprehensive utility-management strategy, DSM has generally been feasible whenever its implementation cost has been lower than the cost of new power supply. However, because improving energy efficiency, and thus reducing the amount of electricity sold, is usually contradictory to the business interests of electricity supply utilities,17 DSM programs have generally been successful in only two cases: (a) where the utilities are relatively responsive to public sector mandates and (b) when energy-efficiency efforts are promoted in combination with measures such as power-factor correction and load-management efforts that are clearly in the utilities' financial interests (Taylor and others 2008).

Standards and labeling. In developing and transition economies, programs to enhance energy efficiency across all sectors have primarily focused on the establishment of energy-performance labels for motors and other mass-produced equipment, certification of energy managers and auditors, energy audits of government buildings, and assistance to industry in energy-use benchmarking. For setting appliance labels and standards, the strategies typically consist of minimum consumption standards for specific equipment; prohibition of manufacture, sale, and import of equipment not conforming to standards; and mandatory labeling to enable consumers to make informed choices.

Appliance labeling is a common instrument for mainstreaming energy efficiency in the household sector in Organisation for Economic Co-operation and Development (OECD) and many middle-income countries, because it allows buyers to take into account not only the initial cost of the appliance, but also the otherwise invisible factor of appliance energy consumption. The range of appliances labeled across countries varies—typically including refrigerators, freezers, and air conditioners—as do the types of labels used. The endorsement label (used in Brazil and currently under consideration in Uruguay), typically defined as voluntary, is used to indicate products that belong to the most energy-efficient class or meet predetermined standard or eligibility criteria. Comparative labels, by contrast, are primarily mandatory and allow customers to judge the energy efficiency and relative ranking of all products that carry the label. In Latin America and the Caribbean, this type of label is common in Colombia, Costa Rica, and Mexico (Harrington and Damnics 2004).

Industrial sector end-use efficiency. Energy use in the industrial sector accounts for about 40 percent of the world's electricity consumption. Industrial energy efficiency thus offers an important opportunity for developing countries with expanding industrial infrastructure to increase their competitiveness by adopting best energy-efficiency practices from the outset in new industrial facilities. Because about 65 percent of electricity in the industrial sector is being consumed by electric motor systems, energy-efficiency improvements in this area have large potential for energy savings. For example, the use of variable-speed drives and efficient pumps, motors, compressors, and fans presents an energy savings potential of about 40 percent. In the aluminum industry, which involves the energy-intensive production of aluminum from bauxite, large energy savings—as much as 90 percent—can be obtained by making aluminum from recycled products (Greenpeace International and EREC 2007).

Much like commercial and residential buildings, industrial facilities have rarely achieved energy efficiency through the competitive pressures of the marketplace alone. However, unlike commercial and residential buildings, the presence in industrial facilities of individual energy-efficient components, such as pumps, boilers, and compressors, does not ensure that entire industrial systems will be energy-efficient, with major losses occurring because of equipment misapplications and the energy-conversion process.

An alternative solution is a policy of voluntary commitments for energy-efficiency improvement by industry, in which the government and industry agree to negotiated targets for up to 10 years, allowing for planning and implementation of strategic energy-efficiency investments. Other alternatives include industrial energy-management standards and requirements for companies to set energy-efficiency goals and to adopt appropriate practices. Such mandatory standards for energy management are currently in place in countries such as Denmark, Ireland, and Sweden. In the future, the dissemination of energy-efficient industrial technologies to other countries can be expected to accelerate through international standardization of testing procedures and norms pursued by the International Organization for Standardization.

Residential sector end-use efficiency. The various energy and electricity end uses in the residential (and buildings) sector can be ranked according to importance in terms of their share in total energy use as well as according to their potential for energy-efficiency improvements. According to estimates for OECD economies (IEA 2008), space heating is by far the most important energy use in the residential sector, account-ing for 53 percent of household energy end use in 2005, followed by appliances (21 percent), water heating (16 percent), and lighting and cooking (5 percent each).

Global experience shows that very little of the efficiency potential in the residential sector has so far been captured and that it is unlikely to be tapped based on market-based incentives alone. Some of the reasons that households do not undertake "profitable" energy-efficiency mea-sures include (a) a lack of information on the benefits of energy-saving lights and appliances, (b) principal-agent problems, in which the benefi-ciaries (residents) of energy-efficiency improvements are different from those who make the investments (landlords); (c) the higher up-front cost of more energy-efficient equipment and the lack of financing of such equipment; (d) the fact that building designers and developers underin-vest in energy-efficient designs and systems to lower "first cost," which raises life-cycle costs to occupants; and (e) electricity subsidies to resi-dential customers that hinder investments in efficient building retrofits and end-use technologies.

Additional barriers to demand-side energy efficiency, applicable not only to the residential sector, but also across all consumer groups and common in Latin America and the Caribbean, include (a) limited techni-cal and risk-management skills in the energy-efficiency field; (b) limited incentives for power distribution companies to promote decreased elec-tricity consumption; and (c) inadequate or absent policy or regulatory incentives for energy efficiency, including rigid procurement policies and regulatory frameworks that fail to allow utilities to finance investments in energy efficiency by allowing customers to repay through their electricity bills.

Efficiency improvements typically are also inhibited by a lack of access to commercial financing for energy-efficiency projects. Although the energy service company (ESCO)[18] industry has developed in many industrial countries for the purpose of financing energy-efficiency investments, there is a general absence or underdevelopment of ESCOs in developing and middle-income countries. ESCOs have not arisen to the same extent in developing countries in part because of the legal and contract-intensive nature of the ESCO business. Where ESCOs do exist, they often have limited access to capital and, thus, operate as fee-for-service energy-efficiency consultants rather than finance the investments themselves.

End-use efficiency potential. A number of industries in Latin America and the Caribbean have significant potential for further efficiency improvements. For instance, in Brazil's chemical and petrochemical industry, the International Energy Agency estimates a potential energy-efficiency improvement of 21 percent (IEA 2008). In Mexico, significant energy savings in the industrial sector could be achieved through cogeneration, with an estimated 85 percent of the potential not yet used. Cogeneration could provide as much as 12.5 percent of new electricity capacity in Mexico, at costs that are significantly lower than the country's current marginal costs of power generation (Johnson and others 2010).

According to the same study, measures to improve the efficiency of industrial motors alone could result in electricity savings of about 114 TWh by 2030, approximately equivalent to the amount of electricity consumed by Mexico's industrial sector in one year. If one extrapolates the assumptions about the available savings from improvements in industrial motors to other countries in Latin America and the Caribbean, more than 500 TWh of electricity could be saved by 2030, or about 22 TWh per year.

In Mexico's residential sector, several large-scale energy-efficiency projects have been implemented as part of a program under the state utility CFE, managed by the Trust Fund for Electricity Savings (Fideicomiso para el Ahorro de Energía Eléctrica, or FIDE). As a result of the program's first phase alone (2002–06), about 25,000 homes were insulated and 623,000 refrigerators and 130,000 air-conditioning units replaced, with associated electricity savings of 2.1 TWh. In its second phase, planned for 2009–12, the program is expected to generate electricity savings of as much as 13.5 TWh (World Bank 2009b), equivalent to the annual combined consumption of the Dominican Republic and Uruguay.

Mexico's new Special Program for Climate Change (Programa Especial de Cambio Climático, or PECC, 2009–12) identifies several high-impact energy-efficiency measures that could, in the near future, be cost-effectively implemented in the residential, commercial, and industrial sectors. Measures specific to the residential sector include the replacement of refrigerators and air-conditioning equipment as well as thermal insulation of buildings, estimated to result in electricity savings of 7.4 TWh over the period 2009–12. The application of similar energy-efficiency tools in selected commercial and municipal buildings is expected to save 2.1 TWh over the same period.

Over the longer term, as much as 200 TWh could be saved by 2030 through investments in residential air conditioning, lighting, and refrigeration. This amount is more than four times Mexico's current annual electricity consumption in the residential sector. If the same efficiency potential is assumed for other countries in Latin America and the Caribbean, the electricity savings by 2030 in the residential sector alone could amount to as much as 1,000 TWh, comparable to the combined annual electricity demand for Brazil and Mexico.

In Argentina, studies indicate significant untapped potential for energy savings across the economy, particularly in the industrial sector. If one assumes reasonable rates of market penetration of energy-efficiency technologies and practices, electricity savings of about 20 percent could be achieved, while 30 percent of electricity could be saved in the commercial sector through improved efficiency in lighting and air conditioning (World Bank 2008).[19] In the country's residential sector, replacement of inefficient lamps and appliances could help reduce residential electricity consumption by up to 30 percent, and similar electricity savings, in percentage terms, are feasible through upgrades in public lighting.

Supportive Energy-Efficiency Policies

The experiences of Latin America and the Caribbean show that positive changes in energy efficiency have generally been driven by a combination of factors, including higher energy prices, better design or organizational measures and technical improvements, new technologies, energy-conservation programs, and competition. Demand-side efficiency programs have generally been more successful when (a) electricity tariffs reflect market values (opportunity costs and the long-run marginal cost of supply); (b) legislation and regulatory policies promote energy efficiency, such as the enforcement of sound environmental, building, and appliance standards; and (c) fiscal policies penalize the production and import of energy-inefficient technologies and reward energy efficiency.

Dedicated energy-efficiency and energy-conservation policies have yielded sizeable results and, from the viewpoint of costs, can be seen as an alternative source of energy for the future as the region faces the need to respond to growing electricity demand. In the future, further energy-efficiency gains could be achieved through improvements in product manufacturing, as well as through social education and information campaigns promoting corporate social responsibility and addressing the importance of energy savings that are already playing an important role in the European Union and other industrial countries.

The key remaining barriers to the market penetration of energy-efficient technologies include (a) the absence of financial intermediation by lending institutions to develop energy-efficiency lending, (b) weak incentives for building efficient new buildings as a result of the principal-agent problem, and (c) a relative paucity of ESCOs and other private sector energy-efficiency service delivery mechanisms. As in other regions, in Latin America and the Caribbean there is a need to more consistently pursue efficiency investments by electricity and gas distributors, with each rewarded for saving either form of energy.

Further improvements in power sector efficiency are also dependent on such factors as the definition of comprehensive national strategies for the overall energy sector; policy and financial support for the modernization of electricity grids and transmission and distribution infrastructure; and a realignment of utility regulations and rate structures to provide utilities with incentives for efficiency rather than increased generation. Other factors that can promote energy efficiency include the following:

- Establishment of national energy-efficiency labeling requirements, like those already in place in Argentina, Brazil, Chile, Colombia, Costa Rica, Cuba, Jamaica, Mexico, and Peru, and further efforts toward creating a *harmonized system* of standards and labeling programs, such as those currently pursued by the Mercosur Standards Organization (Asociación Mercosur de Normalización, or AMN) and the Pan-American Standards Commission (Comisión Panamericana de Normas Técnicas, or COPANT)

• Introduction of national certification schemes for electric and electronic equipment in countries where they are absent (for example, the Dominican Republic)
• Design of well-targeted promotion programs (including mass procurement to lower costs and subsidy schemes for low-income consumers) for compact fluorescent lamps and other energy-efficient equipment
• Establishment of an appropriate policy and financing framework (for example, through the establishment of loan guarantee facilities by local commercial banks and the introduction of a utility "public wire charge" as in Brazil) to ensure the economic sustainability of the desired market transformations
• Introduction of customized financial products by commercial banks to match the characteristics of energy-efficiency projects
• Mainstreaming of energy efficiency in government procurement strategies, as is already being done in Mexico and Peru, to ensure that only high-efficiency equipment is being purchased (Aita 2008; Expert Group on Energy Efficiency 2007)
• Introduction of tax incentives to accelerate demand for efficient technologies and services
• Development of appropriate legal norms and sanctions to prevent illegal manufacturing and distribution of inefficient or uncertified appliances.

Aggregate Energy-Efficiency Potential

If countries in Latin America and the Caribbean could reduce distributional electricity losses to the region-best of about 6 percent—the level of Chile—the Dominican Republic, Nicaragua, and Paraguay could obtain electricity savings of above 20 percent of the current total annual generation (figure 4.5). The combined annual savings for the region from reaching Chile's benchmark would amount to about 78 TWh—comparable to the combined annual amount of electricity sold in all of Central America, Chile, and Peru, or equivalent to about one-fourth of the annual electricity sales in Brazil. Although this may be an extremely optimistic scenario

Figure 4.5   Annual Electricity Savings from Reducing Distributional Losses to Chile's Level

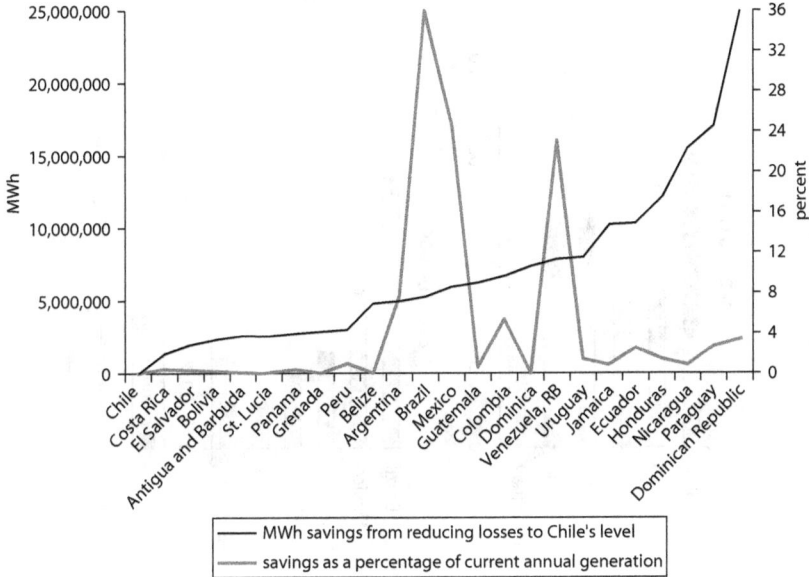

Source: Authors' calculations based on Benchmarking Data of the Electricity Distribution Sector in the Latin America and the Caribbean region 1995–2005 database, World Bank.

for distribution loss reduction, it is conservative because no savings are assumed for improved supply efficiency.

Using Argentina's estimates of end-use energy-efficiency potential in the industrial, residential, and commercial sectors and extrapolating them across the Latin America and the Caribbean region results in annual electricity savings of 230 TWh (using 2008 electricity consumption figures). The savings range from 42 GWh in Grenada to 95 TWh in Brazil (figure 4.6).

If one uses the estimates of the supply- and demand-side energy-efficiency potential outlined previously, it is possible to compare these against the results of the modeling exercise presented in chapter 3. The supply-side measures could reduce energy demand by 78 TWh, while the demand-side measures were estimated to be 230 TWh (figure 4.7). Together, these rather simple estimates show that overall demand could be reduced by about 12 percent.

Figure 4.6   Extrapolation of Argentina's Efficiency Level, Annual Savings

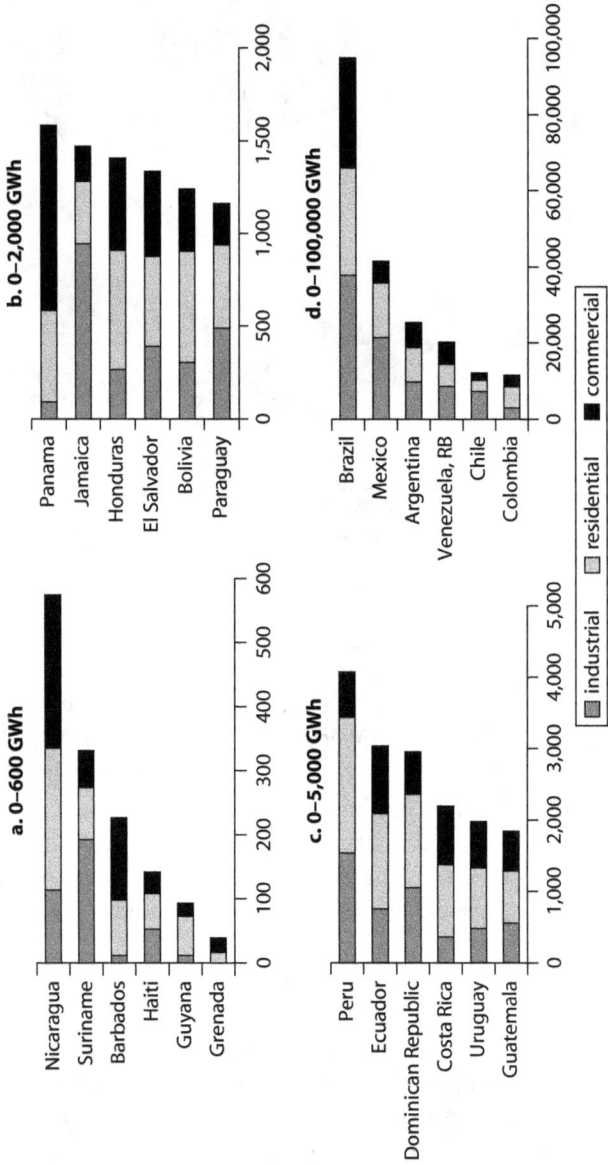

a. 0–600 GWh

b. 0–2,000 GWh

c. 0–5,000 GWh

d. 0–100,000 GWh

■ industrial   □ residential   ■ commercial

Source: Authors' calculations based on Benchmarking Data of the Electricity Distribution Sector in the Latin America and the Caribbean region 1995–2005 database, World Bank.
Note: GWh = gigawatt-hour.

Figure 4.7   Electricity Demand Scenario with Energy Efficiency Potential Realized

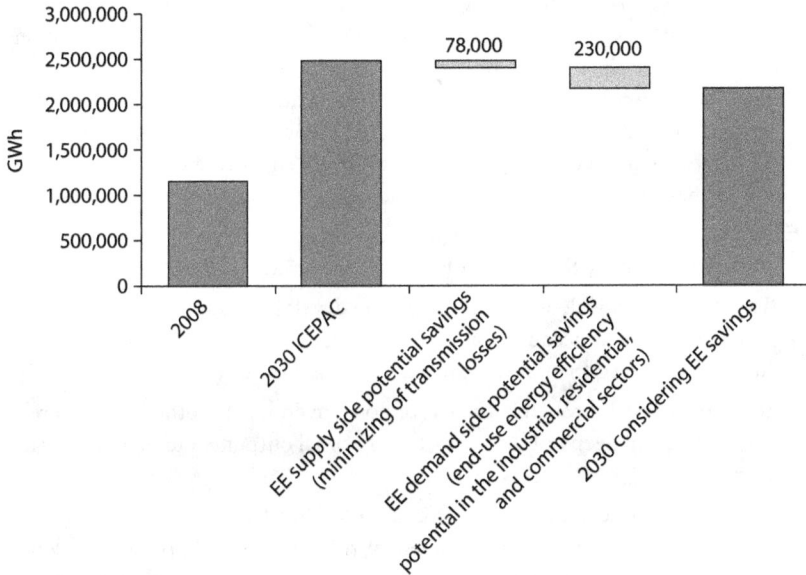

Source: Authors' elaboration based on optimization model.
Note: EE = energy efficiency, GWh = gigawatt hour.

Notes

1. In addition to hydroelectricity, Brazil uses a large amount of sugarcane bagasse for heat and electricity generation, much of it used by the sugar industry itself. During the 30-year period between 1975 and 2005, the contribution of sugarcane bagasse to Brazil's final energy consumption increased from 4 percent to 11 percent. Bagasse's 11 percent share of total energy production in 2005 compares to the 4 percent of energy provided by ethanol (Ministério de Minas e Energia, "Brazil's Final Energy Consumption." http://www.mme.gov.br).

2. At least for the time being, solar power's importance for electric power generation in Latin America and Caribbean lies in the use of solar photovoltaic panels, primarily for off-grid and rural applications. Although photovoltaic systems serve a vital social and development purpose, their total contribution to power generation will be limited. As such, solar is not addressed extensively in this report. Nonetheless, solar hot water systems and passive solar designs, though not directly contributing to electricity generation, are universally large in the region and could offset large amounts of electricity.

3. Reference is made here to the Battelle Wind Power Classification, which is widely used in wind power mapping. The average wind speed cited is measured at a height of 50 meters.

4. This amount equals Brazilian R$148.33 per MWh at R$1.80 per US$1.00.

5. In Denmark, maximum wind generation can provide more than 100 percent of the country's power generation, demonstrating that previous assumptions about limiting the amount of wind in the system are proving not to be true. Spain has also installed a large percentage of wind capacity, and has not experienced difficulties resulting from intermittency. The fact that European countries are interconnected, and that there is a significant capacity of hydropower, has contributed to the absorption of wind and other renewables.

6. However, this latter breakthrough would seem unlikely to occur in time to substantially affect the output of biomass electricity before 2025.

7. This value is equivalent to what might be developed in the United States by 2015, adjusted to account for land area and a heat flux that is 25 percent higher on average than in the United States and Canada.

8. If there is a breakthrough in the use of nonconventional geothermal resources, especially from deep hot dry rocks, the potential output could be substantially increased.

9. Both the Mexico and Brazil low-carbon studies have estimated very large potential for residential and commercial solar hot-water applications, which in Mexico have costs that are competitive with electric and liquified petroleum gas (LPG) hot-water systems. In the Mexico analysis, a large-scale solar water-heating program was estimated to be able to reduce carbon dioxide emissions by about 18.9 million tons per year, which is equivalent to the reduction of about 23 TWh of electricity.

10. This range assumes sugarcane production of about 1.2 billion tons in 2030 in Latin America as a whole. If there were a large expansion of ethanol production, sugarcane production would be substantially higher and, hence, the associated range of electricity generated. For example, in the World Bank's (2010) *Brazil Low-carbon Country Case Study*, discussed in the section titled "A Closer Look at Hydropower," the Bank presents a scenario of 1.73 billion tons of sugarcane in Brazil alone (a generation of 200 TWh).

11. For example, in Brazil, small hydro (greater than 30 MW) technical potential was estimated to be about 17.5 GW (EPE 2007), compared with 252 GW of larger hydro resources. Brazil's definition of "small hydro" is larger than in most countries. Note also that small hydro plants tend to have a lower capacity factor than larger plants (that is, fewer kilowatt-hours of output per kilowatt of capacity).

12. As a part of the Initiative for the Integration of South American Infrastructure, there is a large hydroelectric project under construction. The Madeira River project is an initiative to integrate Bolivia, Brazil, and Peru. The project consists of two hydroelectric dams: Santo Antonio (installed generating capacity of 3,150 MW) and Jirau (installed capacity of 3,300 MW).

13. The differences are even more dramatic if Brazil's imports of Paraguay's share of Itaipu are taken into account.

14. Mexico has had a connection with Belize for a number of years, providing small (from Mexico's perspective) but important (from Belize's perspective) amounts of power. The fact that Belize already benefits from the interconnection with Mexico is one of the main reasons that Belize is the only country in Central America that is not part of the SIEPAC system.

15. For comparison, load management changes only the time when the energy is consumed (Lovins 2005).

16. Jamaica's Ministry of Energy projects a reduction in total losses to about 18 percent by 2020, compared with 24 percent in 2008 (Watson 2009).

17. Among the exceptions are when DSM programs help reduce peak-load requirements, when DSM reduces consumption among nonpaying customers, and when there is a shortage of new power capacity and the DSM program can free up power to allow the connection of new customers.

18. The so-called "energy service company" represents a range of company types that invest in energy efficiency, often providing the upfront financing and entering into "guaranteed savings" contracts with industrial or commercial customers. Such contracts provide the consumer with a guarantee that their energy bills will be lower by an agreed-to amount after the investment. The ESCO uses the energy savings to recoup the investment. See Taylor and others 2008.

19. According to the electricity consumption figures for 2008, these combined potential electricity savings in Argentina's industrial and commercial sectors are roughly equivalent to the entire amount of electricity consumed in 2008 in Bolivia, Costa Rica, and Guatemala combined (OLADE 2009).

References

Aita. 2008. "Política Energética Nacional: Planificación Energetica." Prospectiva Energetica al 2032 Taller Subregional de America del Sur, Ministerio de Energia y Minas, Peru.

BEN (Balanço Energético Nacional). 2008. Empresa de Planejamento Energética, Rio de Janeiro.

CIER (Comisión de Integración Energética Regional). 2008. "Proyecto CIER 15 Fase II Informe Final." CIER, Montevideo, Uruguay.

Devoto, G. A. 2007. "Hydroelectric Power and Development in Argentina." Ente Nacional Regulador de la Electricidad, Buenos Aires.

EERE/USDOE (Energy Efficiency and Renewable Energy Department/U.S. Department of Energy). 2008. *20% Wind Energy by 2030: Increasing Wind*

Energy's Contribution to U.S. Electricity Supply. Report DOE/GO-102008-2567. Washington, DC: EERE/USDOE.

EPE (Empresa de Pesquisa Energetica). 2007. "Plano Nacional de Energia 2030." Ministerio de Minas e Energia, Brasil.

Expert Group on Energy Efficiency. 2007. *Realizing the Potential of Energy Efficiency: Targets, Policies, and Measures for G8 Countries.* United Nations Foundation, Washington, D.C.

Greenpeace International and EREC (European Renewable Energy Council). 2007. *Energy [R]evolution: A Sustainable World Energy Outlook.* Amsterdam: Greenpeace; Brussels: EREC.

Harrington and Damnics. 2004. "Energy Labeling and Standards Programs throughout the World." Report for the National Appliance and Equipment Energy Efficiency Committee, Australia, July.

IEA (International Energy Agency). 2008. *Worldwide Trends in Energy Use and Efficiency: Key Insights from IEA Indicator Analysis.* Paris: Organization for Economic Co-operation and Development and IEA.

International Development Research Center. http://www.idrc.ca/en/ev-69633-201_850342-1-IDRC_ADM_INFO.html.

Johnson, T. M., C. Alatorre, Z. Romo, and F. Liu. 2010. *Low-Carbon Development for Mexico.* Washington, DC: World Bank.

Komives, K., T. M. Johnson, J. D. Halpern, J. L. Aburto, and J. R. Scott. 2009. *Residential Electricity Subsidies in Mexico: Exploring Options for Reform and for Enhancing the Impact on the Poor.* Washington, DC: World Bank.

Lovins, A.B. 2005. "Energy End-Use Efficiency." Rocky Mountain Institute, Snowmass, CO. http://www.rmi.org/rmi/Library/E05-16_EnergyEndUse Efficiency.

NAS (National Academy of Sciences), NAE (National Academy of Engineering), and NRC (National Research Council). 2009. *Electricity from Renewable Resources: Status, Prospects, and Impediments.* Washington, DC: National Academies Press.

OLADE. 2008. *Energy Statistics Report 2007.* Quito: OLADE.

———. 2009. *Energy Statistics Report 2008.* Quito: OLADE.

Soreide, T., D. Benitez, and F. Haladner. 2009. "The Politics of Cross-Border Electricity Supply: Some Experiences from Latin America." Background Paper, World Bank, Washington, DC.

Taylor, R. P., C. Govindarajalu, J. Levin, A. S. Meyer, and W. A. Ward. 2008. *Financing Energy Efficiency: Lessons from Brazil, China, India, and Beyond.* Washington, DC: World Bank.

UNIDO (United Nations Industrial Development Organization) and REEEP (Renewable Energy and Energy Efficiency Partnership). 2010. "Sustainable Energy Regulation and Policy-Making for Africa: Demand-Side Management (Module 14)." UNIDO, Vienna.

Watson, C. 2009. "Energy Planning Methodology and Approaches: Jamaica's Experience. Latin America and the Caribbean Energy Forecast—Energy Scenarios at 2032." Presentation at OLADE Subregional Workshop, Ministry of Energy, Jamaica, February 19.

World Bank. 2008. "Argentina Energy Efficiency Project." Project Appraisal Document, World Bank, Washington, DC.

———. 2009a. "Energy in Latin America and the Caribbean: Challenges and Opportunities." Energy Unit, Sustainable Development Department, Latin America and the Caribbean Region.

———. 2009b. "Mexico: Lighting and Appliances Efficiency Project." Concept Note, April 1, World Bank, Washington, DC.

———. 2010. *Brazil Low-carbon Country Case Study*. Washington, DC: World Bank.

CHAPTER 5

Conclusions

This report contributes to the discussion of future power supply in Latin America and the Caribbean by (a) supporting the integration of a large database on electricity production and consumption for the majority of countries in the region and developing a methodology for assessing future electricity supply using a consistent framework across countries (chapter 3); and (b) examining a range of promising alternatives for meeting future electricity demand requirements (chapter 4).

Baseline Electricity Supply Scenario

Under quite modest gross domestic product (GDP) growth assumptions—essentially 3 percent per annum for 2014–30—the demand for electricity would more than double from 1,150 terawatt-hours (TWh) to about 2,500 TWh. Under a baseline scenario, Latin America and the Caribbean would need to add more than 239 gigawatts (GW) of new power generating capacity over the next 20 years, much of it from thermal generating capacity. Under a higher income growth scenario, generating capacity would need to grow even more. Regardless of the exact rate of growth of GDP and the ultimate demand for electricity, it is clear that the countries of the region must plan for a significant expansion of electricity generating capacity.

Under a baseline scenario for the region for 2030—aggregating and extrapolating current country power expansion plans—the majority of new generating capacity would be met by hydropower (36 percent) and natural gas (35 percent). In many ways, this is a "best case" scenario, because the current country expansion plans for hydropower- and natural gas–based generation are quite optimistic. The amount of new hydropower that would be required under the baseline scenario is more than 85 GW. This compares to 76 GW of hydropower capacity that was commissioned in the region over the past 20 years, and many of the best sites in terms of power capacity, financial returns, and environmental and social risks have already been exploited. To tap the region's extensive hydro resources, most countries will need to make changes in their regulatory policies.

A large increase in the use of natural gas power generation is also envisioned under the baseline, growing from the current capacity of 60 GW to more than 144 GW in 2030. Compared to petroleum and coal, the growth of natural gas for power generation would have both positive efficiency benefits through the use of combined-cycle technology and positive environmental benefits by the reduction in local pollutants and lower carbon emissions. Realizing the expansion of natural gas–fired generating capacity will require more regional cooperation in building pipelines and negotiating bilateral gas contracts, because natural gas resources are not evenly distributed throughout the region. Some countries, such as Mexico, will need to expand domestic gas production and to rely on increased trade to meet the aggressive expansion plans for natural gas–fired power generation. In some countries, low prices for domestically produced natural gas have led to the inefficient use of natural gas, which could be overcome by a combination of pricing reforms and technology standards. For example, by converting "open-cycle" natural gas plants to more efficient combined-cycle technology, Peru could add 800–900 megawatt (MW) of power capacity, equivalent to about two years of new capacity additions at current rates of growth.

Alternatives for Meeting Latin America and the Caribbean's Power Demand

In addition to hydropower and natural gas, there are a number of other options for meeting the electricity demand needs of the region. However, because of data limitations and the fact that most of the alternatives explored do not feature prominently in current country power expansion plans, they were not addressed in the modeling exercise in chapter 3. Among the findings and preliminary conclusions of "alternatives" from chapter 4 are the following:

- **Non-hydro renewables**—especially wind—could provide an important new source of electricity in most countries of the region, which would help to diversify the overall electricity supply mix. Wind and other renewables have been aided considerably over the past five years by significant reductions in technology supply costs and by the prospects of a growing carbon market. As noted earlier, hydropower is by far the most important renewable energy option for the region, but the supply of hydropower (and natural gas) can be complemented by significant contributions of non-hydro renewables to meet growing demand.

- **Increased electricity trade** could provide significant new capacity by enlarging the regional electricity market and lowering overall supply costs in the process. Although the region is poised to make greater use of trade for supplying electricity demand, there remain obstacles, both regulatory and political, that have inhibited trade in the past but that can be overcome through more concerted regional actions.

- **Greater energy efficiency** can help reduce electricity demand and increase the effective electricity supply in a cost-effective manner. Major potential exists for improving the efficiency of electricity supply, including reducing transmission and distribution losses and tapping the large amount of cogeneration potential in industry. There is probably even greater potential to reduce energy demand by improving end-use efficiency, with significant potential in all major sectors. As elsewhere, the lack of a supportive regulatory and policy framework makes investments in energy efficiency less attractive than building new power capacity, but the costs to society of not taking advantage of efficiency potential are large.

Although difficult to quantify, these alternatives could significantly reduce the amount of new thermal generating capacity that would be needed in Latin America and the Caribbean over the next 20 years. Perhaps more important, these alternatives can help to produce a more diversified and stable power sector, and many options have lower costs than traditional power generation solutions. Initial and partial estimates of the contribution of alternatives suggest the following: (a) actively promoting a program of non-hydro renewables, including wind, biomass, and geothermal, could provide between 15 and 30 percent of the total electricity supply by 2030; (b) expanded electricity trade is likely to lower costs by allowing the development of larger-scale projects and also reduce the need for reserve capacity;[1] and (c) regional electricity demand could be reduced by 10–15 percent through limited supply-side and demand-side energy-efficiency measures at costs that are less than those of building new power generation capacity. The overall results are shown in table 5.1 and figure 5.1.

Recommendations[2]

This report focuses on long-term electric power needs and supply options for Latin America and the Caribbean. Although an extensive analysis of the policy and institutional issues confronting electricity power development in the region is beyond the scope of this report, a number of policies are needed to allow the region to meet its growing electricity needs in an efficient, diversified, and environmentally sustainable manner. Developing institutional capabilities and tools for analyzing electricity supply needs for individual countries and the region would also help improve the development of the power sector.

Table 5.1   Electricity Demand and Supply in Latin America and the Caribbean, 2008 and Various ICEPAC Scenarios
TWh

	2008	ICEPAC	ICEPAC-REEF1	ICEPAC-REEF2
Demand	1,153	2,479	2,171	2,171
Efficiency	0	0	308	308
Supply side	0	0	78	78
Demand side	0	0	230	230
Supply	1,153	2,479	2,171	2,171
Hydropower	675	1,239	1,239	1,239
Thermal	431	1,036	530	215
Other renewables	18	102	300	615
Wind	2	33	220	340
Biomass	5	49	55	150
Geothermal	11	20	25	125
Nuclear	32	102	102	102

Source: Authors' elaboration.
Note: REEF = Renewable Energy and Energy Efficiency alternative scenarios, ICEPAC = Illustrative Country Expansion Plans Adjusted and Constrained.

Figure 5.1   Electricity Supply Mix in Latin America and the Caribbean, 2008 and Various ICEPAC Scenarios

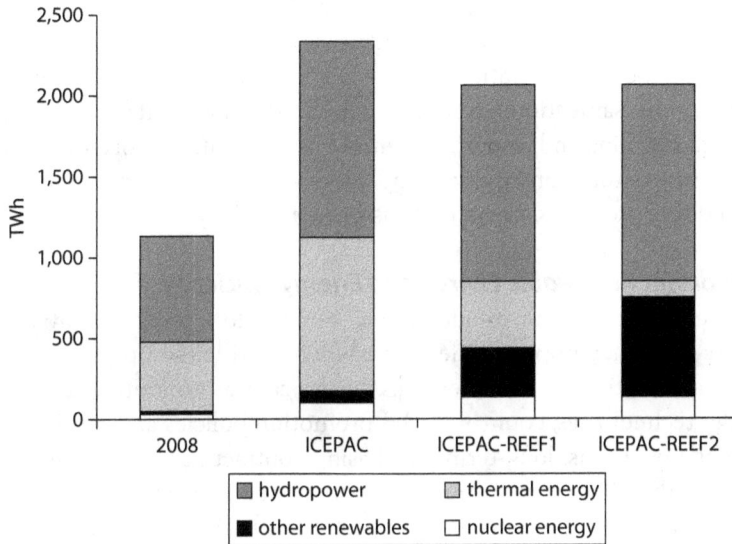

Source: Authors' elaboration based on optimization model.
Note: ICEPAC = Illustrative Country Expansion Plans Adjusted and Constrained, REEF = Renewable Energy and Energy Efficiency alternative scenarios.

Policies and Regulations for Hydropower

The proposed increases in hydroelectric capacity in many countries will require changes in the way power plants have been financed—a greater role for the public sector in regulating and guaranteeing hydroelectricity construction and greater role for the private sector in taking on long-term construction and operation contracts. Hydropower development has also been hindered by the real and perceived social and environmental risks of developing large-scale plants. As such, there is a need to improve the process for identifying and managing social and environmental issues associated with hydropower plants and to improve the environmental consultation, licensing, and commissioning process. The development of hydropower has also suffered from preferential prices for domestic natural gas in several natural gas–producing countries.

Price and Institutional Reform for Natural Gas

Numerous countries in the region have used low preferential pricing for domestic natural gas resources as a way to stimulate natural gas–fired power generation. Although such policies have been partially responsible for the expansion of natural gas–fired power generation in several countries, low prices have also resulted in the inefficient use of natural gas for power generation ("open-cycle" plants) and in inadequate incentives for new natural gas development. Countries that continue to subsidize natural gas for power generation should begin to phase out such subsidies, while at the same time strengthening the development of natural gas through planning and resource inventory assessment, support for pipeline and infrastructure construction, and adoption of transparent and consistent bidding processes for natural gas power capacity.

Support for Renewable Energy and Energy Efficiency

The technical aspects of identifying renewable energy potential and energy-efficiency opportunities must be coupled with appropriate regulatory and institutional frameworks. In the case of promoting renewable energy technologies, countries need promotion policies and mechanisms, such as tax credits, long-term purchasing contracts, and dispatch priorities. Another important aspect for both renewable energy and energy efficiency is the definition of appropriate tariffs that allow producers or consumers to recover the cost of their investments. Several countries in Latin America and the Caribbean have already approved and implemented new laws and regulations that promote energy efficiency and renewable energy use.

Strengthening of Power Sector Planning

Countries in the region should improve their power sector planning. Although the majority of these countries already have power expansion plans, several countries have not yet developed electricity-specific demand and supply growth scenarios. Governments should undertake longer-horizon planning to match power sector investments that are similarly long-term in nature. Among the countries that do have power expansion plans, in several cases the time horizon is limited to 12 years or less, while for others the plans are not updated frequently.[3] Countries also need to be able to develop and discuss with their constituencies realistic power expansion plans that include a wide range of supply options and information on evolving power technologies and international market developments.

Improvement of Regional Power-Planning Tools

There is a need to develop robust and user-friendly regional power-planning tools for individual countries and regional organizations. Such tools would greatly enhance the ability and utility of conducting regional power planning exercises. For the current study, the World Bank relied on externally developed power planning models and software that required substantial amounts of data and technical information. Other power planning tools that are typically used by the World Bank and other institutions include MARKAL and WASP, which are also difficult to use and adjust. The development of more robust power-planning tools would improve the ability of organizations to engage with countries in the region on power planning exercises.

More Reliable Inventory Information

As the demand for new electricity investments increases and regulations and policies are developed for both new and conventional electricity sources, it is critical to improve information on the magnitude, location, and quality of energy resources. There is an especially urgent need to improve information on renewable energy resources, which have been less studied than conventional energy resources. Information is important not only for the identification of individual countries' energy resource potential, but also for the identification of binational or regional energy potential.

Notes

1. The impact of increased trade is complicated to quantify. From the modeling exercise undertaken for Central America, increased trade resulted in the share of hydropower rising from 46 percent to 54 percent, simply by increasing the scale of hydropower plants that could be built in Central America. If trade were expanded beyond Central America, to either Mexico or South America, the gains to trade in terms of lower costs of electricity for Central America could be even larger. From the rising share of hydroelectricity and the consequent reduction in thermal power, carbon dioxide emissions in Central America were found to fall by 14 percent.

2. The policy recommendations presented in this report build upon the lessons learned and issues discussed in several other recent studies by the World Bank on the energy sector, including *Low-Carbon Development for Mexico* (Johnson and others 2010), *Brazil Low-carbon Country Case Study* (World Bank 2010a), "Peru: Overcoming the Barriers to Hydropower" (World Bank 2010b), *Low Carbon Development: Latin American Responses to Climate Change* (de la Torre, Fajnzylber, and Nash 2010), *An Overview on Efficient Practices in Electricity Auctions* (World Bank, forthcoming), and *Peru: Downstream Natural Gas Study* (World Bank, forthcoming).

3. For example, Peru's expansion plan provides projections through 2015, and Panama's through 2014. Although El Salvador's expansion plan provides projections through 2020, it has not been updated since 2003.

References

de la Torre, A., P. Fajnzylber, and J. Nash. 2010. *Low Carbon Development: Latin American Responses to Climate Change.* Washington, DC: World Bank.

Johnson, T. M., C. Alatorre, Z. Romo, and F. Liu. 2010. *Low-Carbon Development for Mexico.* Washington, DC: World Bank.

World Bank. 2010a. *Brazil Low-carbon Country Case Study.* Washington, DC: World Bank.

———. 2010b. "Peru: Overcoming the Barriers to Hydropower." ESMAP Report 53719-PE, Energy Sector Management Assistance Program, Latin America and the Caribbean Region, World Bank, Washington, DC.

———. Forthcoming. *An Overview on Efficient Practices in Electricity Auctions.* Washington, DC: World Bank.

———. Forthcoming. *Peru: Downstream Natural Gas Study.* Washington, DC: World Bank.

Country Data

Figure A.1   Argentina

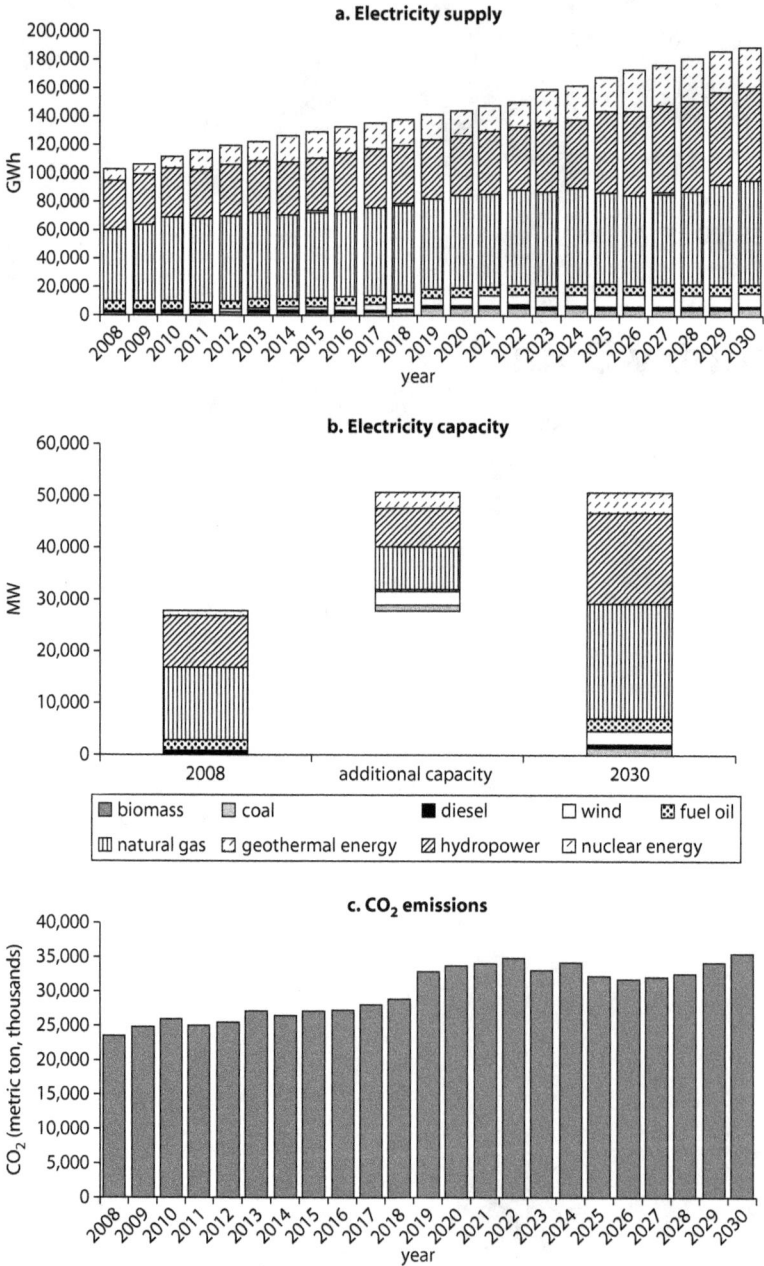

a. Electricity supply

b. Electricity capacity

biomass	coal	diesel	wind	fuel oil
natural gas	geothermal energy	hydropower	nuclear energy	

c. CO₂ emissions

Source: Authors' elaboration based on optimization model.
Note: CO_2 = carbon dioxide, GWh = gigawatt hour, MW = megawatt.

Figure A.2   Bolivia

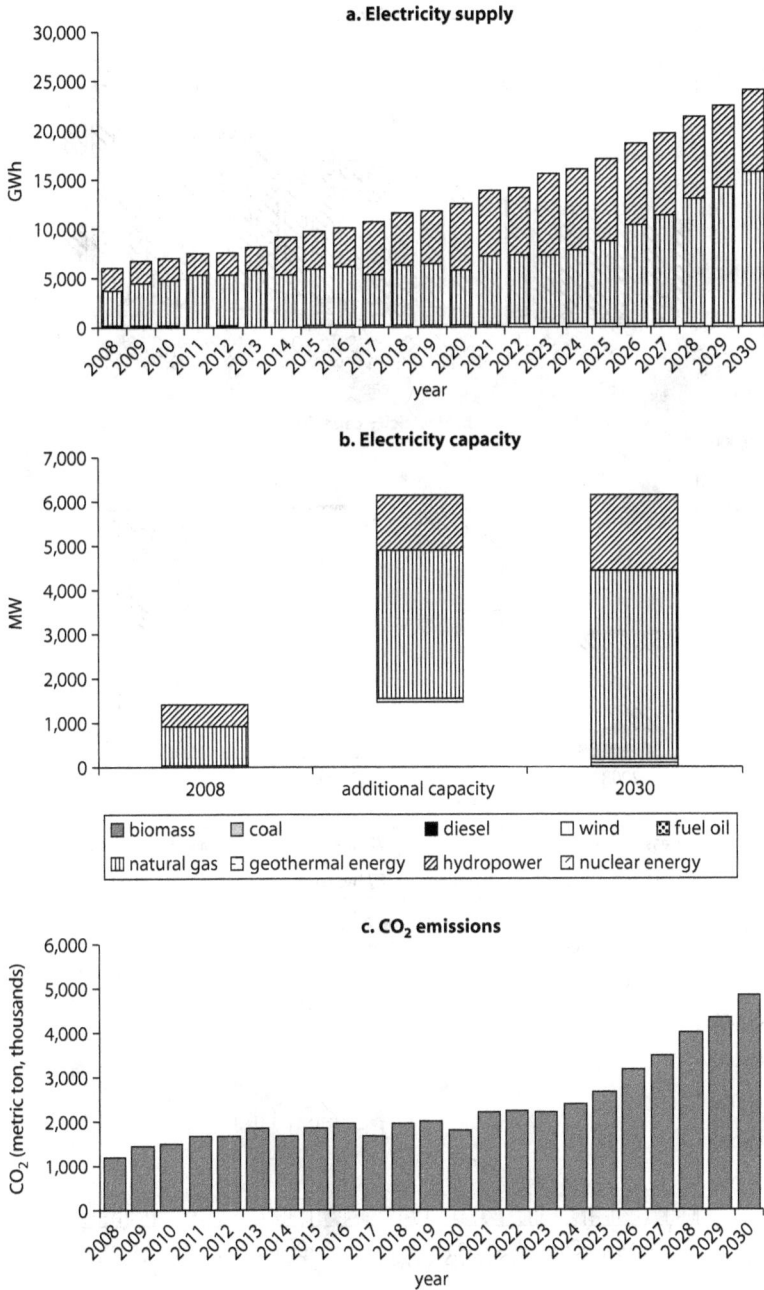

a. Electricity supply

b. Electricity capacity

Legend:
- biomass
- coal
- diesel
- wind
- fuel oil
- natural gas
- geothermal energy
- hydropower
- nuclear energy

c. CO_2 emissions

Source: Authors' elaboration based on optimization model.

Figure A.3   Brazil

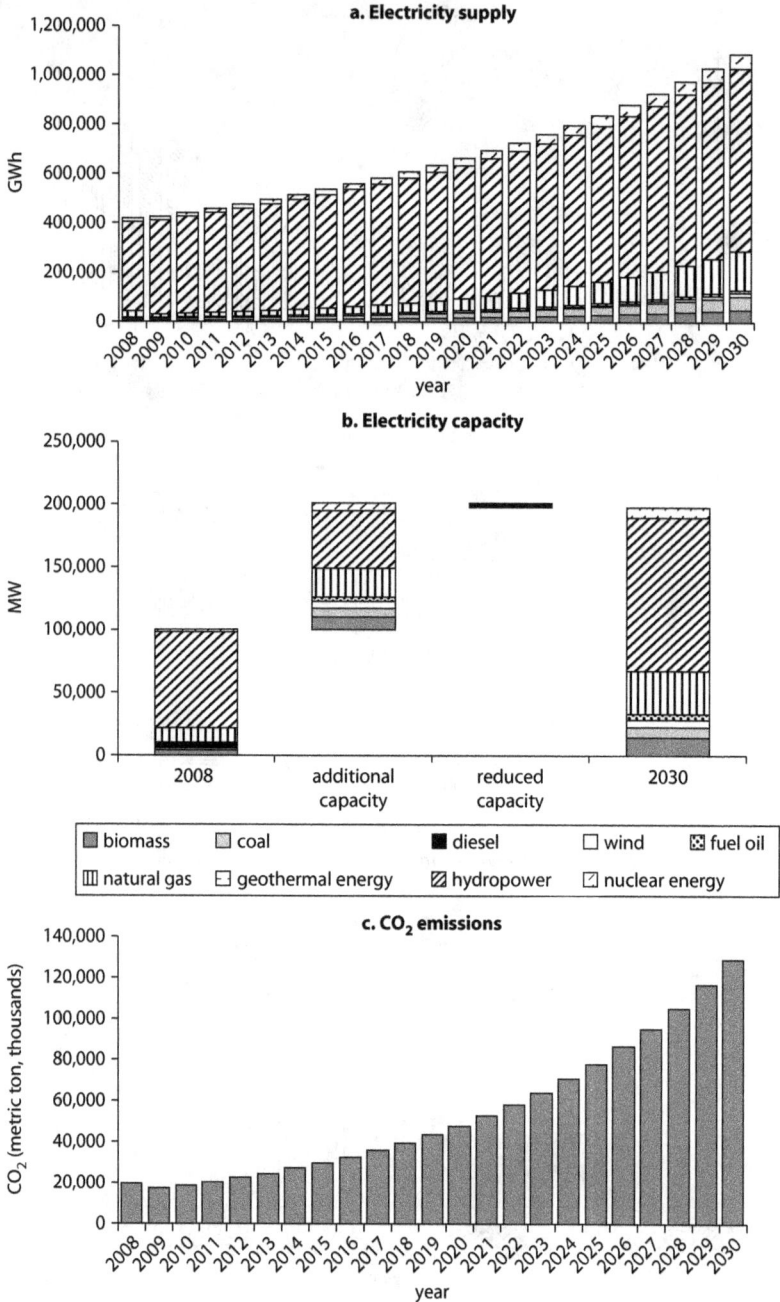

a. Electricity supply

b. Electricity capacity

| biomass | coal | diesel | wind | fuel oil |
| natural gas | geothermal energy | hydropower | nuclear energy |

c. CO_2 emissions

Source: Authors' elaboration based on optimization model.

Figure A.4   Chile

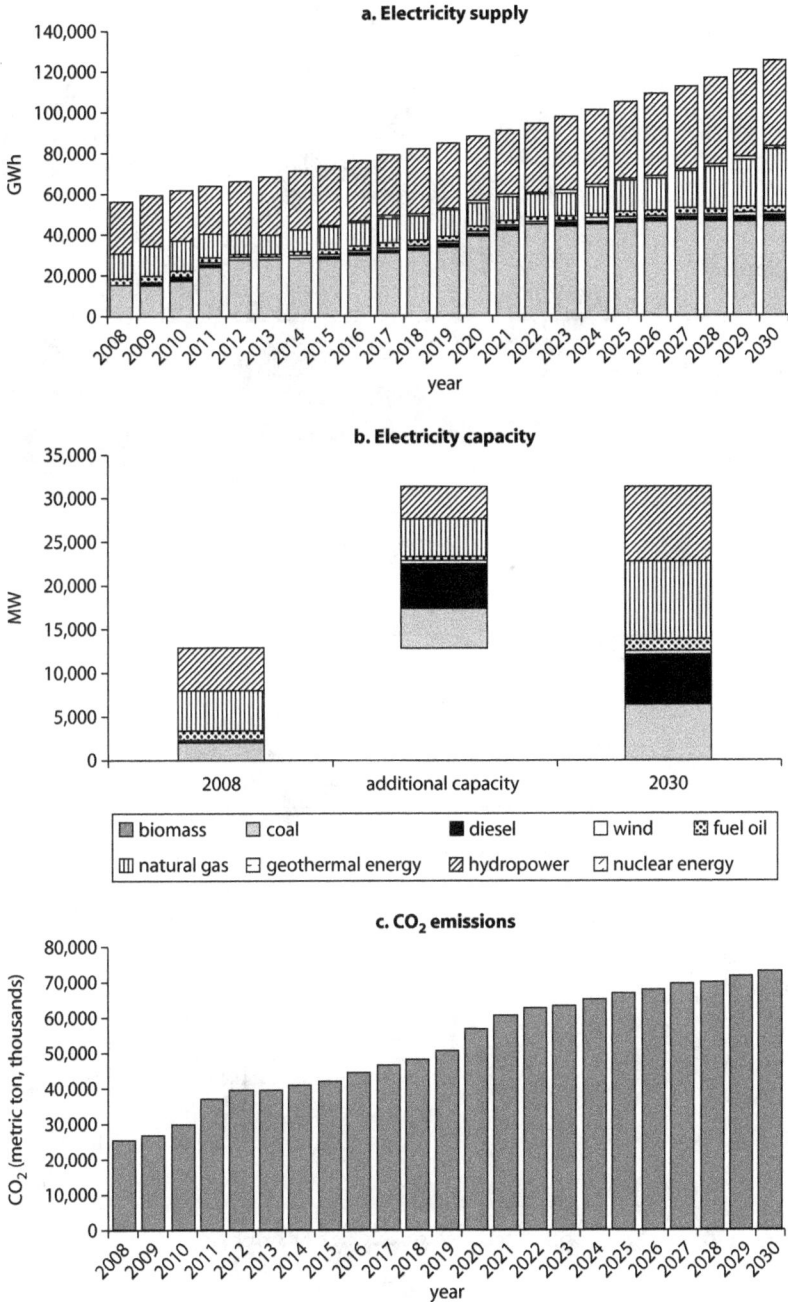

a. Electricity supply

b. Electricity capacity

Legend: biomass · coal · diesel · wind · fuel oil · natural gas · geothermal energy · hydropower · nuclear energy

c. CO$_2$ emissions

Source: Authors' elaboration based on optimization model.

Figure A.5   Colombia

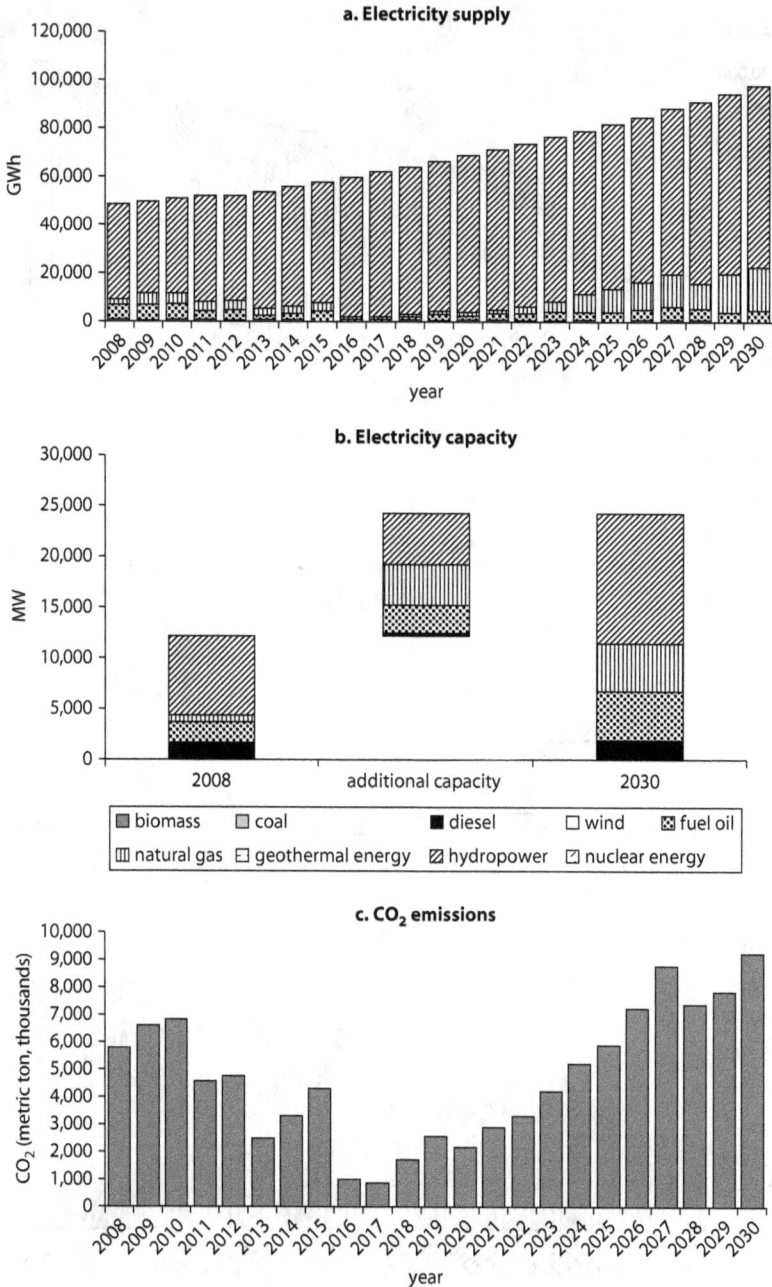

a. Electricity supply

b. Electricity capacity

Legend:
biomass   coal   diesel   wind   fuel oil
natural gas   geothermal energy   hydropower   nuclear energy

c. CO$_2$ emissions

Source: Authors' elaboration based on optimization model.

Figure A.6   Costa Rica

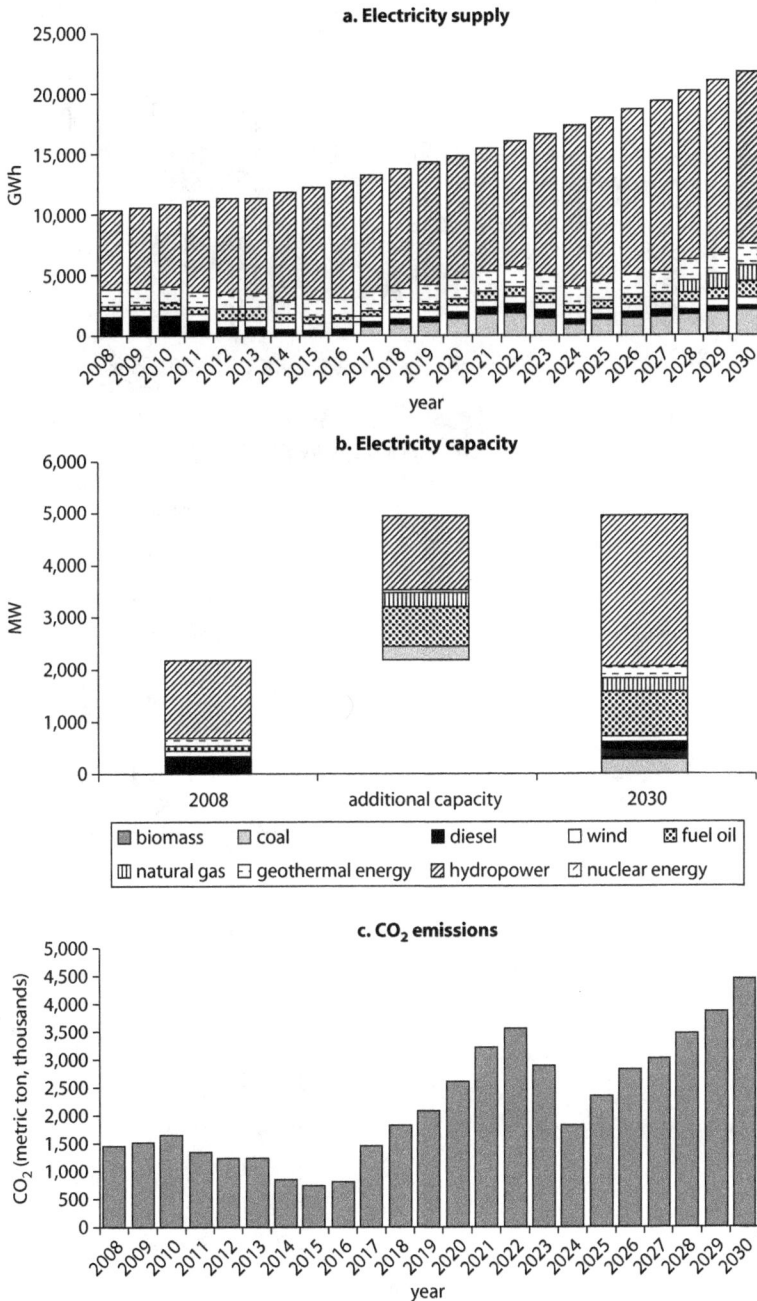

a. Electricity supply

b. Electricity capacity

Legend:
- biomass
- coal
- diesel
- wind
- fuel oil
- natural gas
- geothermal energy
- hydropower
- nuclear energy

c. CO₂ emissions

Source: Authors' elaboration based on optimization model.

Figure A.7   Caribbean

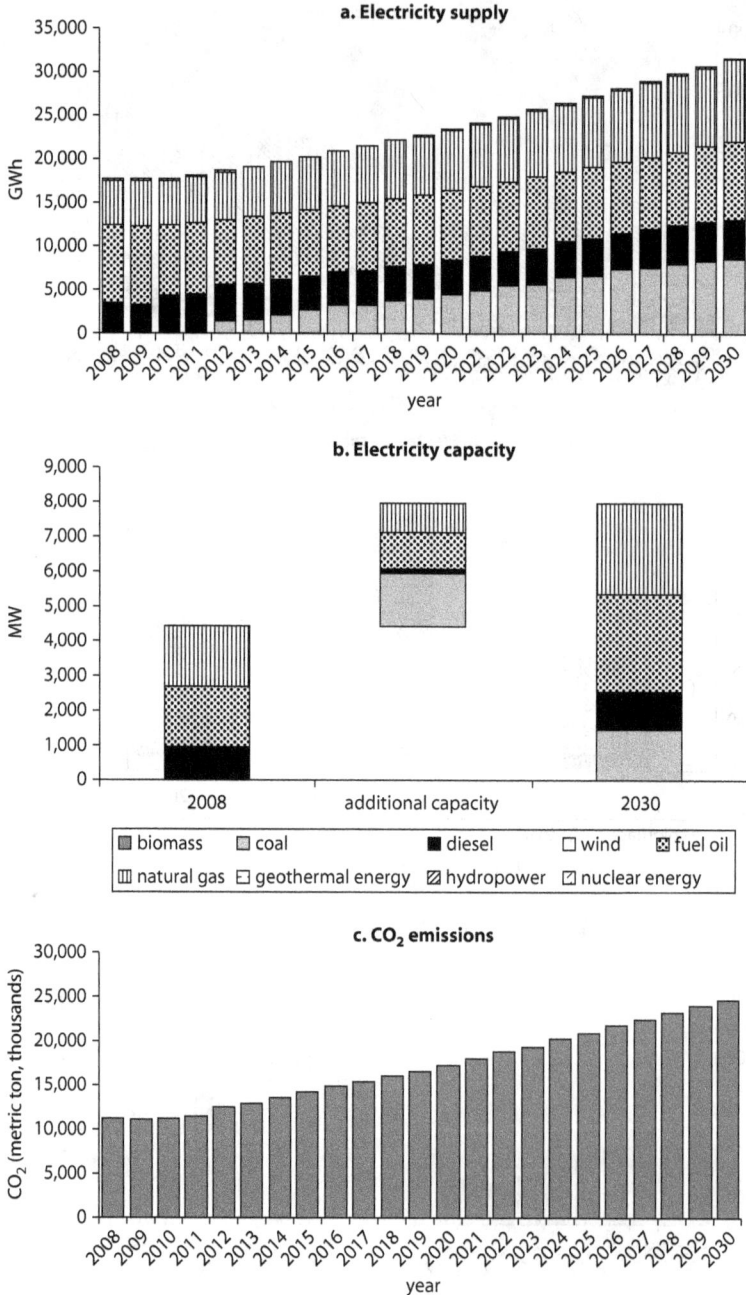

a. Electricity supply

b. Electricity capacity

Legend:
- ▨ biomass   ☐ coal   ■ diesel   ☐ wind   ⊠ fuel oil
- ▥ natural gas   ☐ geothermal energy   ▨ hydropower   ▨ nuclear energy

c. CO$_2$ emissions

Source: Authors' elaboration based on optimization model.

Figure A.8   Ecuador

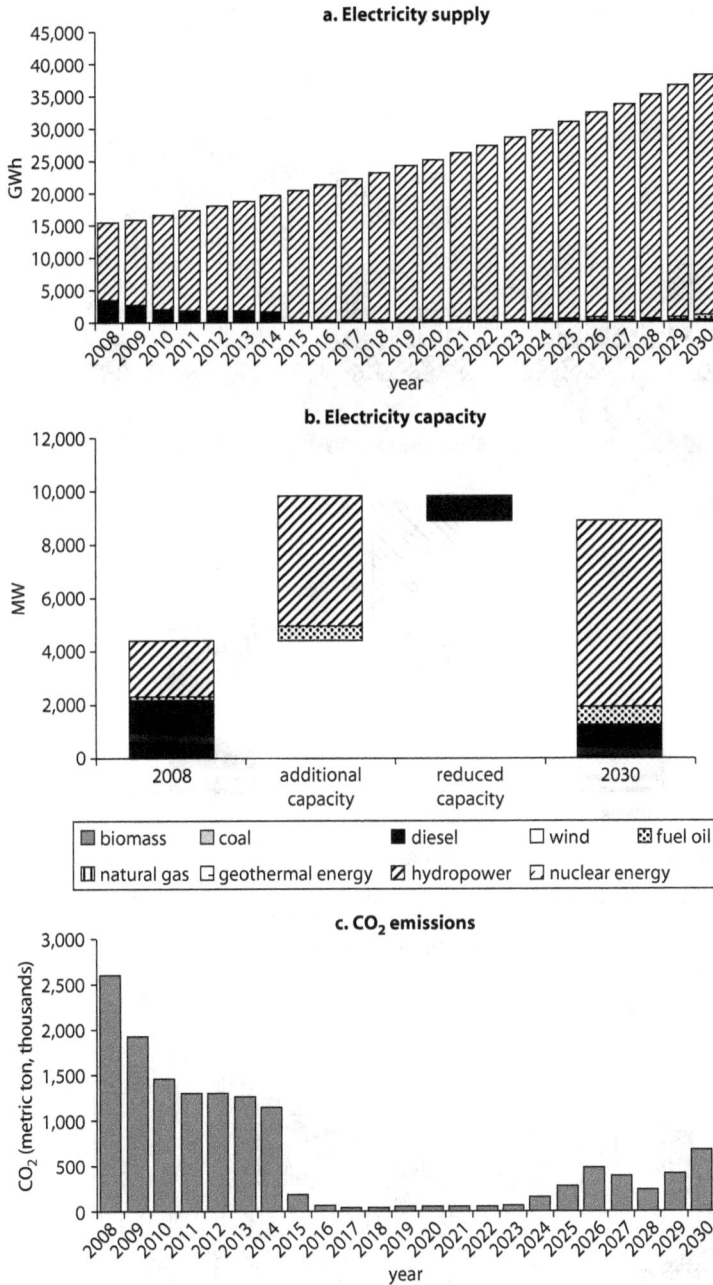

a. Electricity supply

b. Electricity capacity

legend:
biomass   coal   diesel   wind   fuel oil
natural gas   geothermal energy   hydropower   nuclear energy

c. CO_2 emissions

Source: Authors' elaboration based on optimization model.

Figure A.9   Guatemala

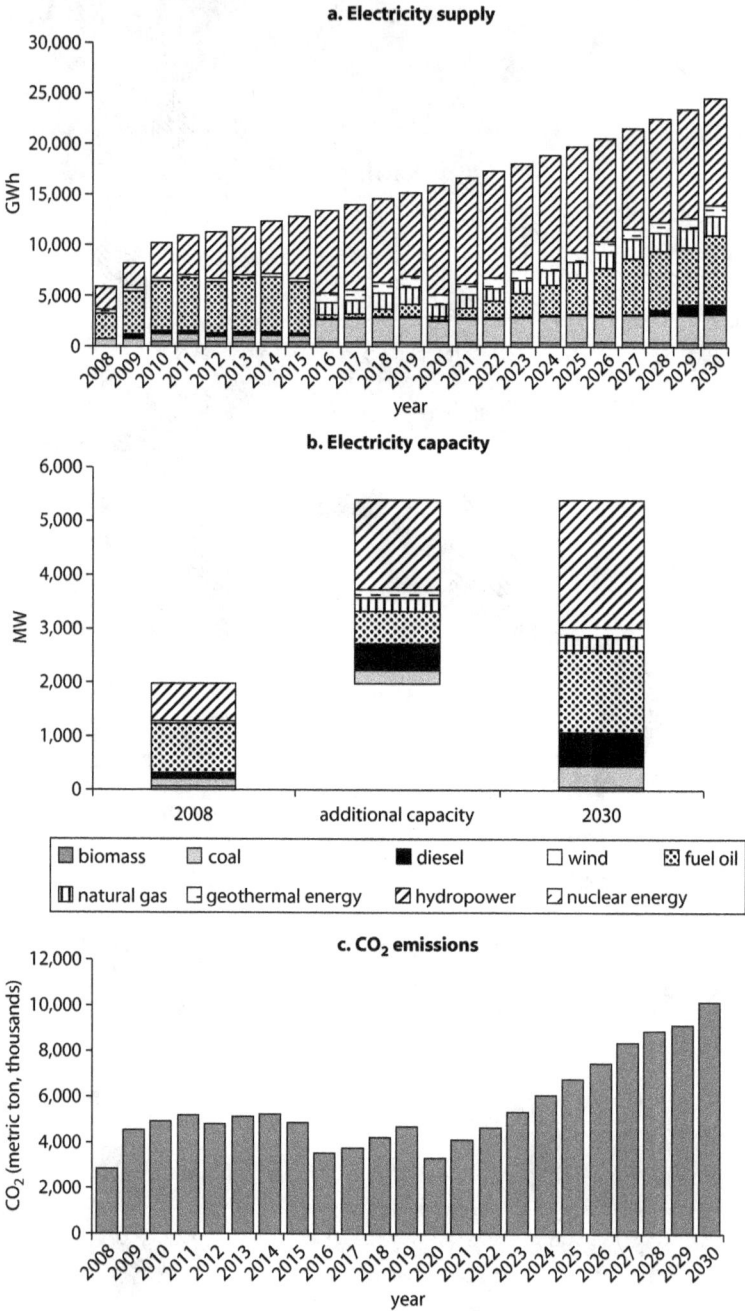

a. Electricity supply

b. Electricity capacity

Legend:
- biomass
- coal
- diesel
- wind
- fuel oil
- natural gas
- geothermal energy
- hydropower
- nuclear energy

c. CO_2 emissions

Source: Authors' elaboration based on optimization model.

Figure A.10   Honduras

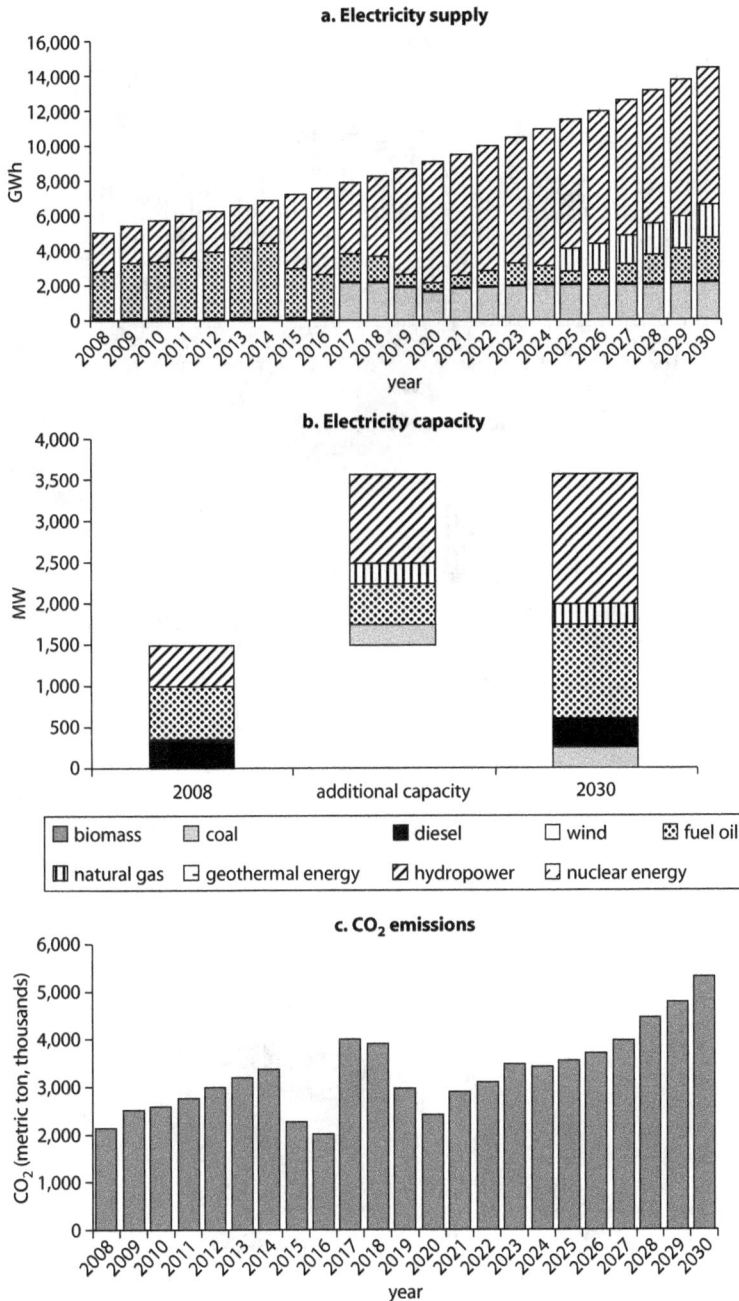

a. Electricity supply

b. Electricity capacity

legend:
| biomass | coal | diesel | wind | fuel oil |
| natural gas | geothermal energy | hydropower | nuclear energy |

c. CO$_2$ emissions

Source: Authors' elaboration based on optimization model.

Figure A.11   Mexico

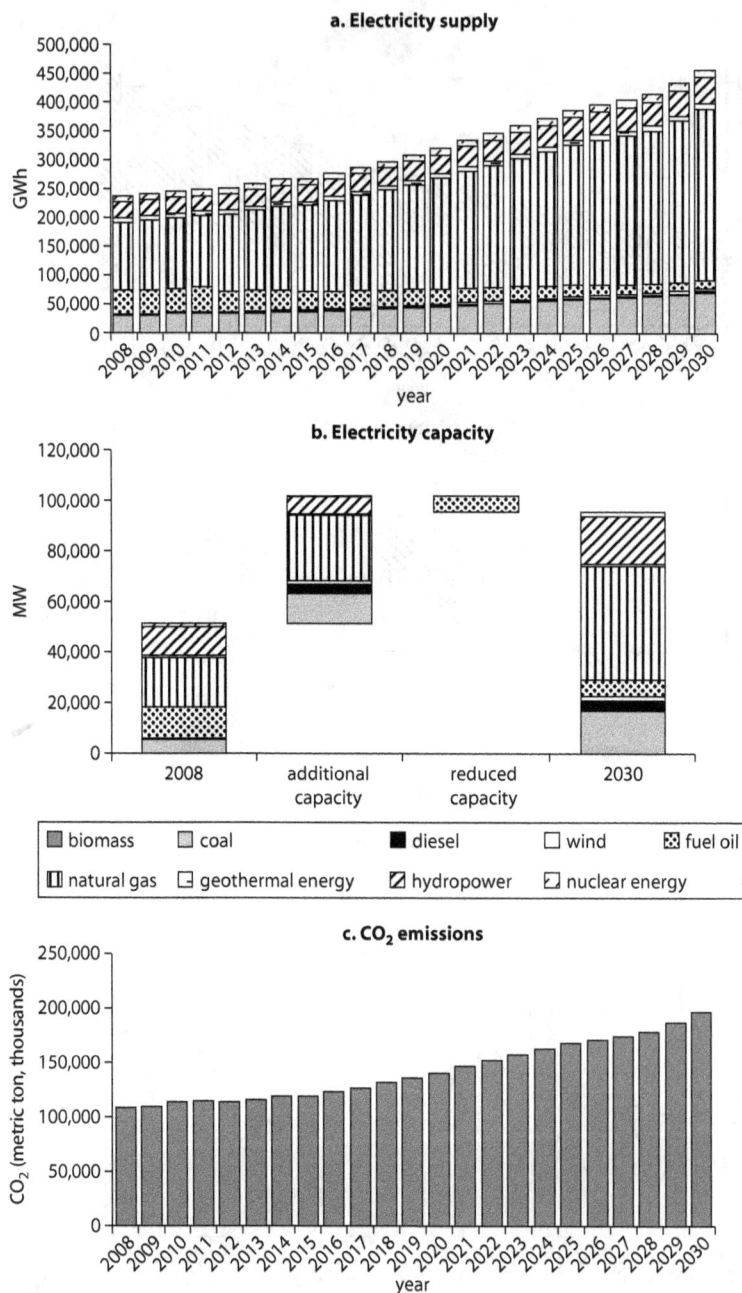

a. Electricity supply

b. Electricity capacity

Legend:
- biomass
- coal
- diesel
- wind
- fuel oil
- natural gas
- geothermal energy
- hydropower
- nuclear energy

c. CO_2 emissions

Source: Authors' elaboration based on optimization model.

Figure A.12   Nicaragua

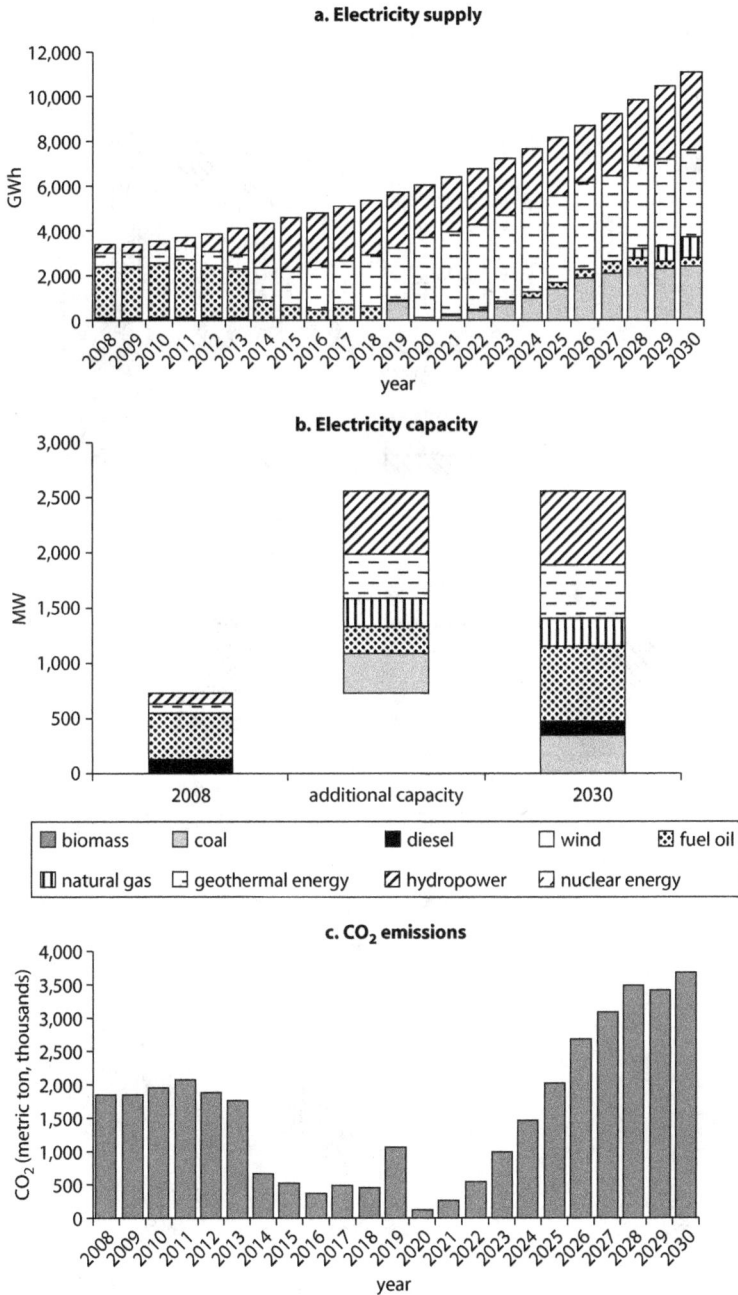

a. Electricity supply

b. Electricity capacity

legend:
- biomass
- coal
- diesel
- wind
- fuel oil
- natural gas
- geothermal energy
- hydropower
- nuclear energy

c. CO_2 emissions

Source: Authors' elaboration based on optimization model.

Figure A.13   Panama

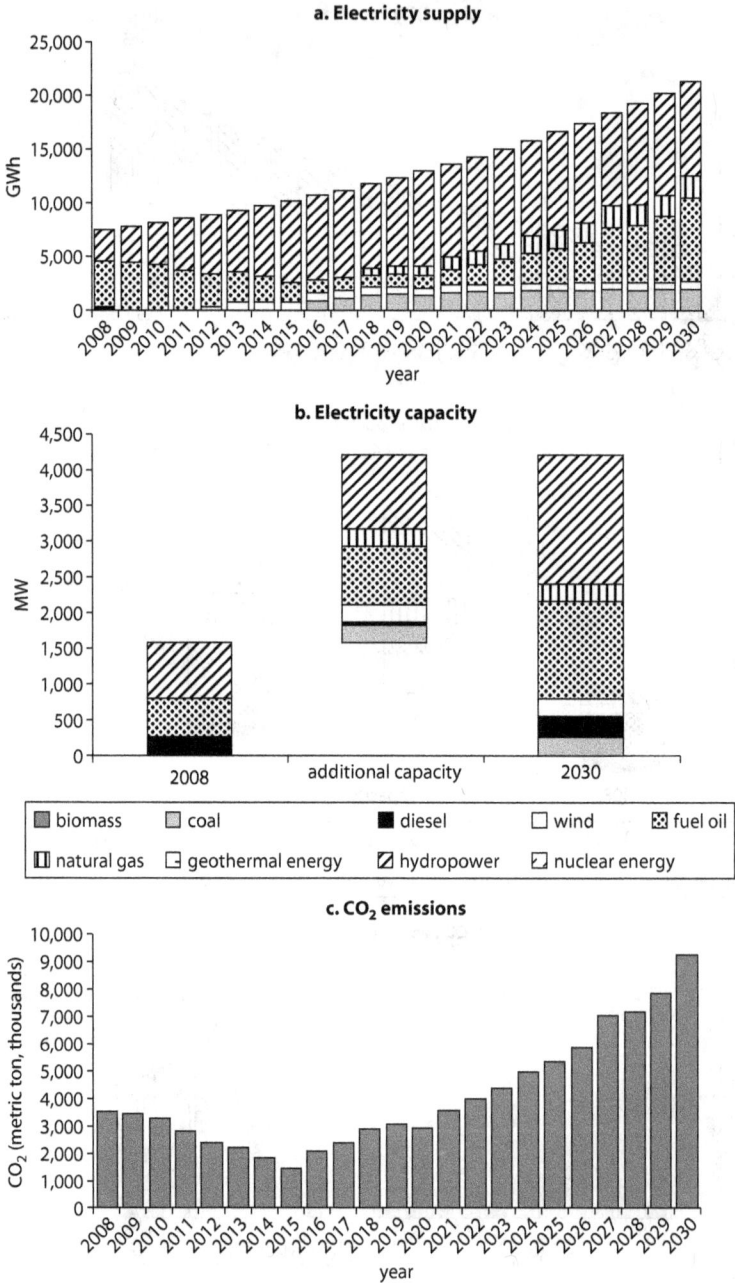

a. Electricity supply

b. Electricity capacity

| biomass | coal | diesel | wind | fuel oil |
| natural gas | geothermal energy | hydropower | nuclear energy | |

c. CO$_2$ emissions

Source: Authors' elaboration based on optimization model.

Figure A.14   Paraguay

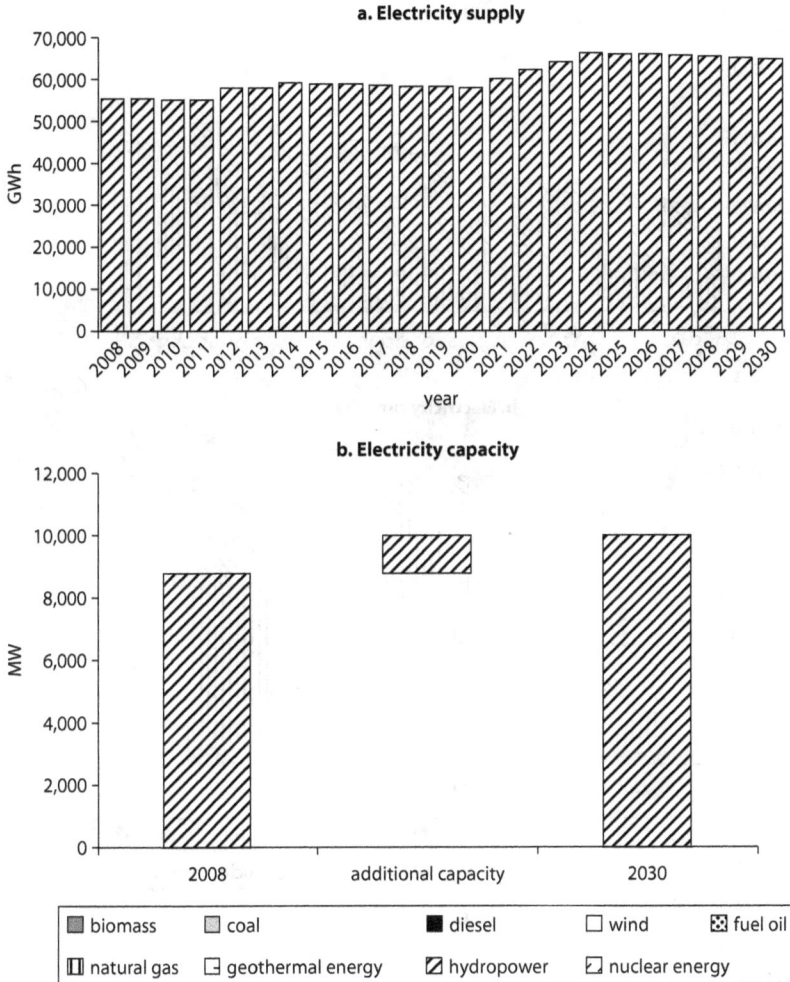

a. Electricity supply

b. Electricity capacity

Legend: biomass, coal, diesel, wind, fuel oil, natural gas, geothermal energy, hydropower, nuclear energy

Source: Authors' elaboration based on optimization model.
Note: CO_2 emissions are zero for every year.

Figure A.15   Peru

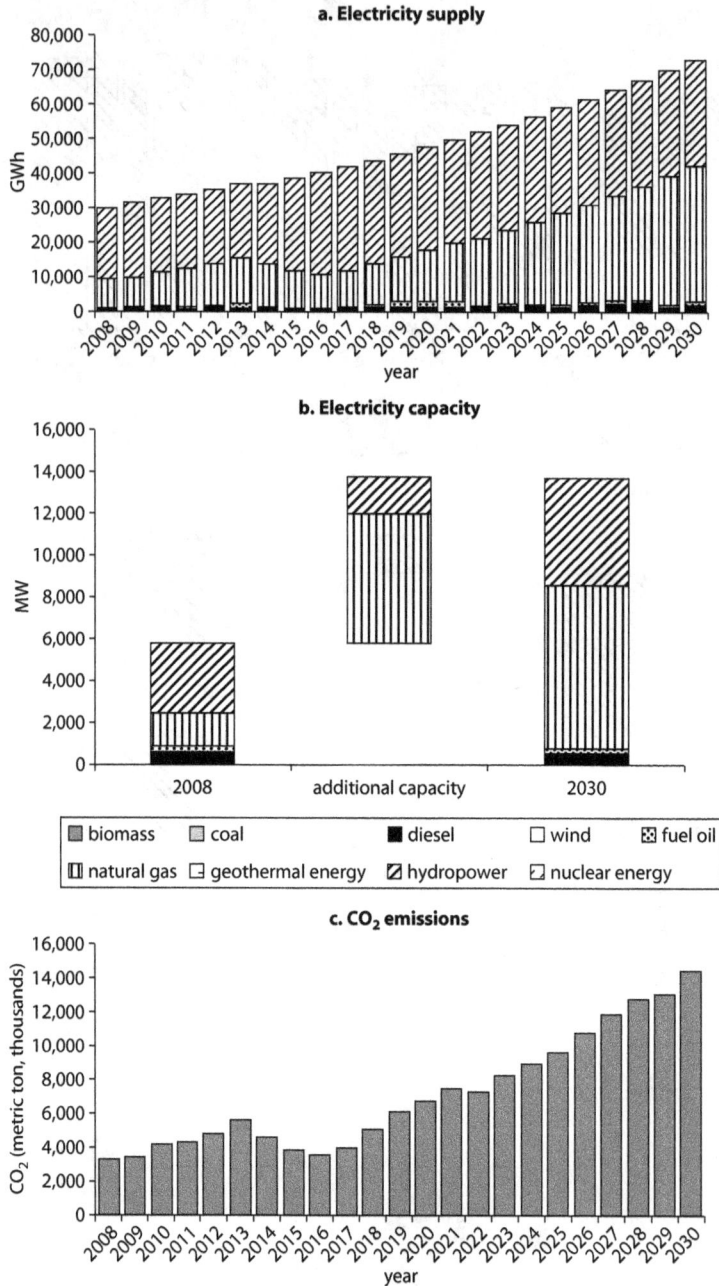

a. Electricity supply

b. Electricity capacity

▦ biomass	▢ coal	■ diesel	☐ wind	⊠ fuel oil
▥ natural gas	▢ geothermal energy	▨ hydropower	▨ nuclear energy	

c. CO$_2$ emissions

Source: Authors' elaboration based on optimization model.

Figure A.16   El Salvador

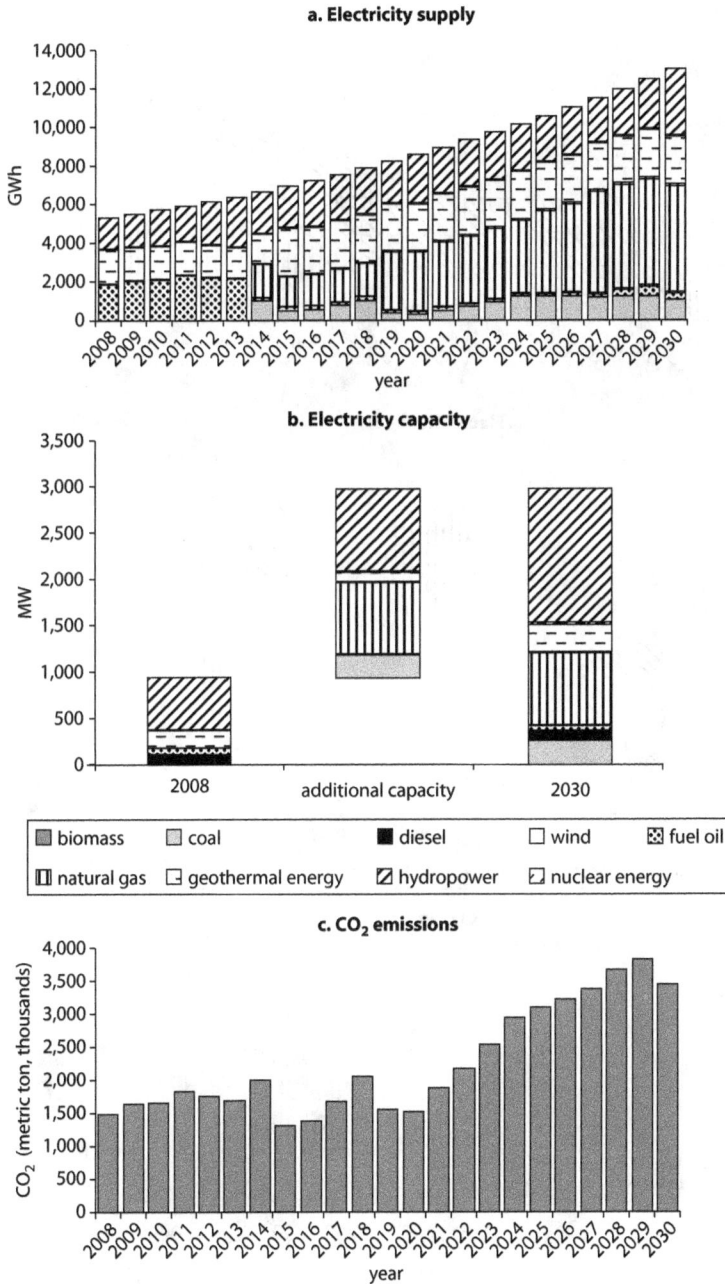

a. Electricity supply

b. Electricity capacity

c. CO$_2$ emissions

Source: Authors' elaboration based on optimization model.

Figure A.17   Uruguay

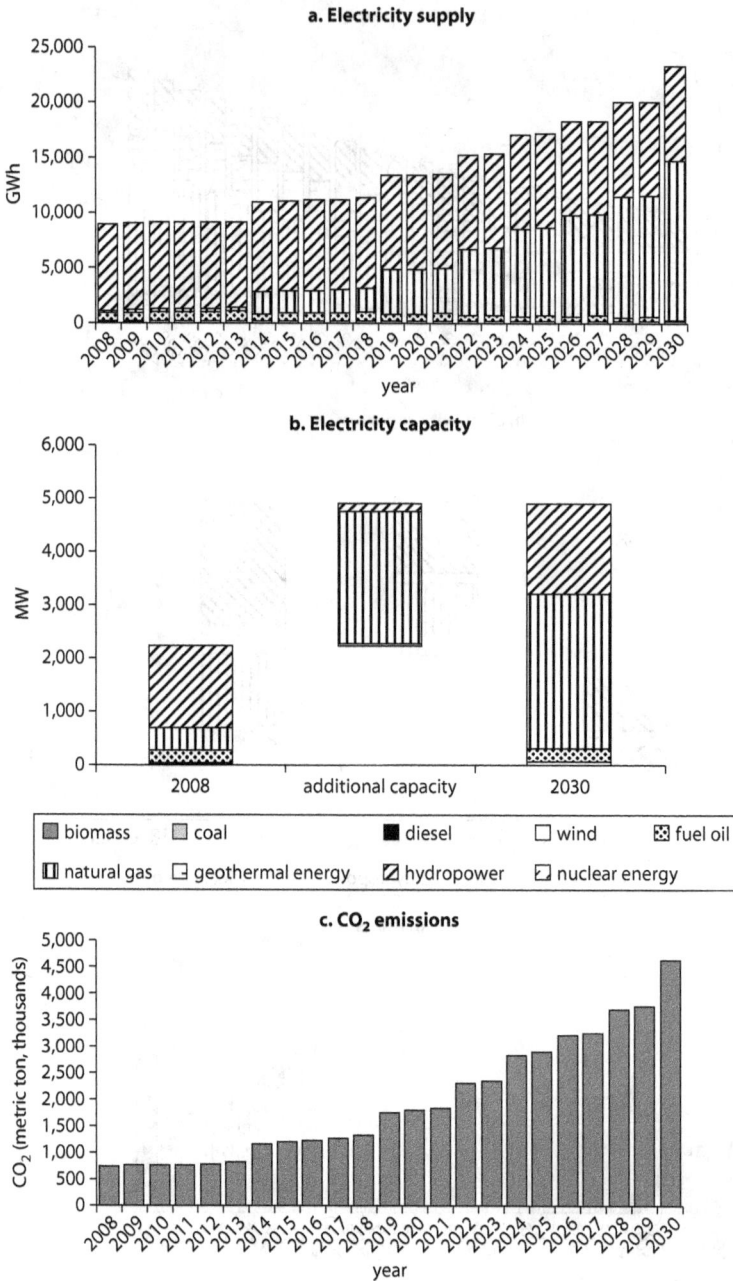

a. Electricity supply

b. Electricity capacity

| biomass | coal | diesel | wind | fuel oil |
| natural gas | geothermal energy | hydropower | nuclear energy | |

c. CO$_2$ emissions

Source: Authors' elaboration based on optimization model.

Price and Income Elasticity of Demand

Similarly to other *normal* goods, the consumption of electricity is expected to increase with a rise in disposable income and the resulting increase in economic activity and purchases of electricity-using appliances, while a rise in electricity prices, ceteris paribus, should lead to a fall in the quantity demanded. Empirical studies focusing on estimating the price and income elasticity of electricity demand generally distinguish between *long-term* elasticities, *short-term* elasticities (one year or less), and time-of-use or *real-time* elasticities. Particularly in academic research, such as that surveyed by Halvorsen (1975), Taylor (1975), and others, there has been explicit analysis of the distinction between long- and short-run effects of price and income. However, in nonacademic research, such as the models of the U.S. Energy Information Administration (1990), the analysis is specific to particular electricity end uses, and the difference between the long-run and the short-run effects is thus generally attributed to the rate of market penetration of various types of housing and appliances. The basic premise of both academic and applied research in specifying the time dimension is that time affects the dependence of capital stock on economic factors, whereby, over the long run, the demand for new housing and intensity of energy use can affect the size of the stock and efficiency of appliances. In the short run, however, the demand for electricity is limited to changes in the use rates given the fixed stock of electricity-using appliances.

Short- and Long-run Price and Income Elasticity of Demand

Caves, Eakin, and Faruqui (2000); Boisvert, Cappers, and Neenan (2002); and Kirschen (2003) have argued that increasing the *short-run* price elasticity of the demand for electrical energy would improve the operation of the market, that is, that significant benefits would accrue if the demand of even a relatively small number of consumers became at least moderately price responsive. The main benefit of this increased overall demand elasticity is an immediate reduction in the magnitude of price spikes, which, in turn, leads to a lower average spot price of electrical energy and, ultimately, affects the price of long-term contracts. Thus, promoting demand responsiveness to price fluctuations can be an efficient instrument for increasing energy supply security, as opposed to simply retaining large amounts of spare generating capacity. Similarly, the specific assumptions about the elasticity of demand for electricity are important in the context of the market power of an electricity provider, namely, the more price elastic the demand, the less market power can be exercised.

In particular, the *real-time* price elasticity of electricity contains important information on the demand response of consumers to the volatility of peak prices. Yet, although excess demand may, in theory, be effectively counteracted by increasing the price responsiveness of demand, in reality, most end users do not observe real-time prices and, hence, cannot react to them. Only a few authors, such as Wolak and Patrick (2001), have explicitly addressed the real-time elasticities, finding fairly low price elasticities—from virtually 0.00 to −0.27[1]—for the five industrial sectors analyzed, although with somewhat higher values at peak demand hours. More recently, Lijesen (2007) analyzes the hour-to-hour price elasticity of electricity demand in the Netherlands, similarly finding a low value for the real-time price elasticity, which the author attributes to the fact that not all users observe the spot market price, because many small consumers are supplied by retailers, often on bilateral contracts with a fixed per unit price.

In addition, according to a number of authors, such as Yusta and Dominguez (2002), Faruqui and George (2002), and Kirschen (2003), who have focused on the short-run as opposed to the real-time elasticity, although demand does decrease in response to a short-term price increase, this effect is relatively small. Presumably, this weak elasticity can be explained by the fact that although the cost of electrical energy represents only a small portion of the total cost of producing most industrial goods or of the cost of living for most households, it is nevertheless indispensable in manufacturing and is regarded as essential to the quality of life by most individuals in industrial societies. As also argued in Heffner and Goldman (2001) and Roos and Lane (1998), even if one were to assume that all consumers are buying electrical energy on the spot market and that they are instantaneously informed of its price, the importance of electricity in daily life represents another barrier to enhancing the elasticity of demand in the immediate term. In the *long run*, this elasticity is typically higher, because consumers have considerably more options, such as switching to gas for heating or purchasing more efficient appliances, or because industrial facilities can relocate to a region with lower average electricity prices. However, as stated by Holtedahl and Joutz (2004), when applied in a developing country context, both short- and long-term energy demand models may require an altogether different framework and interpretation because economic growth and structural change associated with rapid development suggest that income and price elasticities will not be stable.

Most important, however, as summarized in figure B.1, academic research to date shows substantial variation in the estimates of both the price and the income elasticity of electricity demand. For instance, although Elkhafif (1992), Jones (1995), Beenstock, Goldin, and Nabot (1999), Filippini and Pachuari (2002), Urga and Walters (2003), and Holtedahl and Joutz (2004) find short-run price elasticities in the range from −0.04 to −0.18, the estimates found by several other authors, such as Silk and Joutz (1997) and Bjørner and Jensen (2002), are somewhat higher—at about −0.50 or −0.60. As shown by the comprehensive review of econometric studies on the topic by Taylor (1975), the estimates of short-run price elasticity of residential electricity demand typically vary from −0.90 to −0.13, with the long-run price elasticity estimates ranging from −2.00 to 0.00. For commercial electricity demand, the respective values were −0.17 and −1.36. Subsequently, a similar review done by Bohi and Zimmerman (1984) reveals comparable average estimates of short- and long-run price elasticity of residential electricity demand—at about −0.2 and −0.7, respectively, with the results focusing on commercial electricity demand too diverse to arrive at conclusive consensus values. Most recently, Espey and Espey (2004), in their review of 36 studies focusing on residential electricity demand, show the range of short- and long-run price elasticity estimates as between −1.10 and 0.00 and between −2.25 and −0.10, respectively. However, the range of income elasticity estimates is between 0.04 to 0.97 for the short-run models and between 0.20 and 3.74 for the long-run models.

Figure B.1   Comparison of Estimates of Short-run and Long-run Price and Income Elasticity of Demand for Electricity

a. Price elasticity estimates

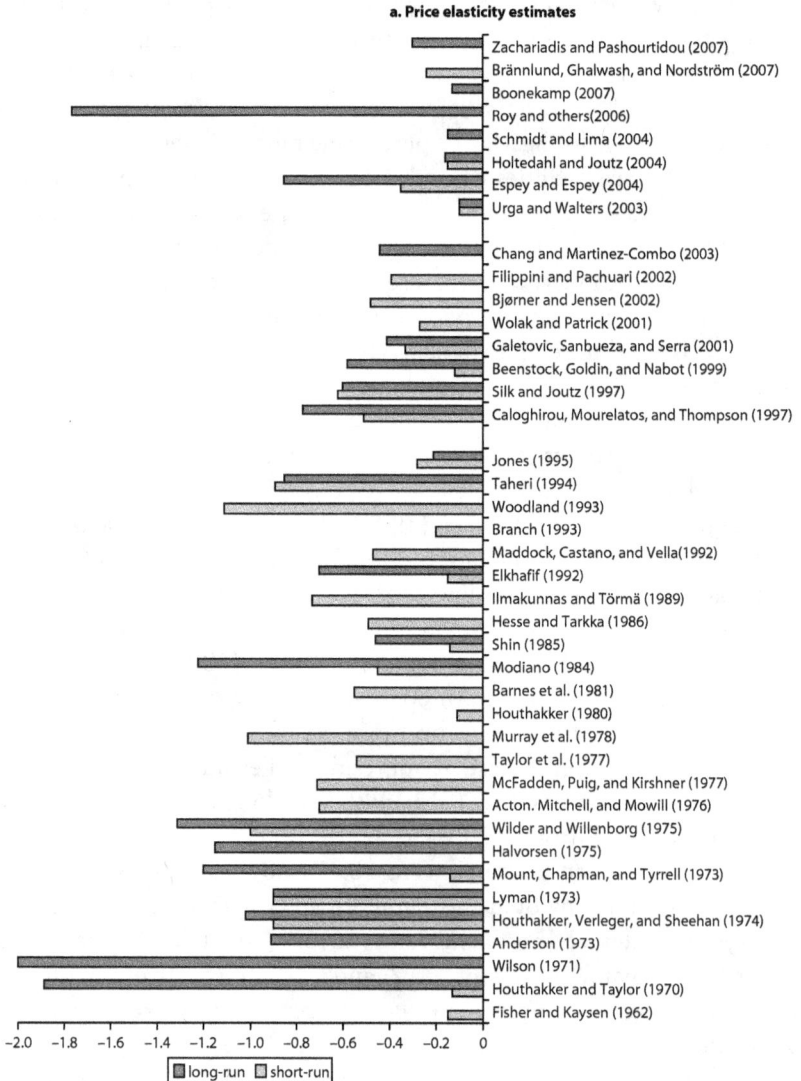

(continued next page)

Figure B.1   *(Continued)*

b. Income elasticity estimates

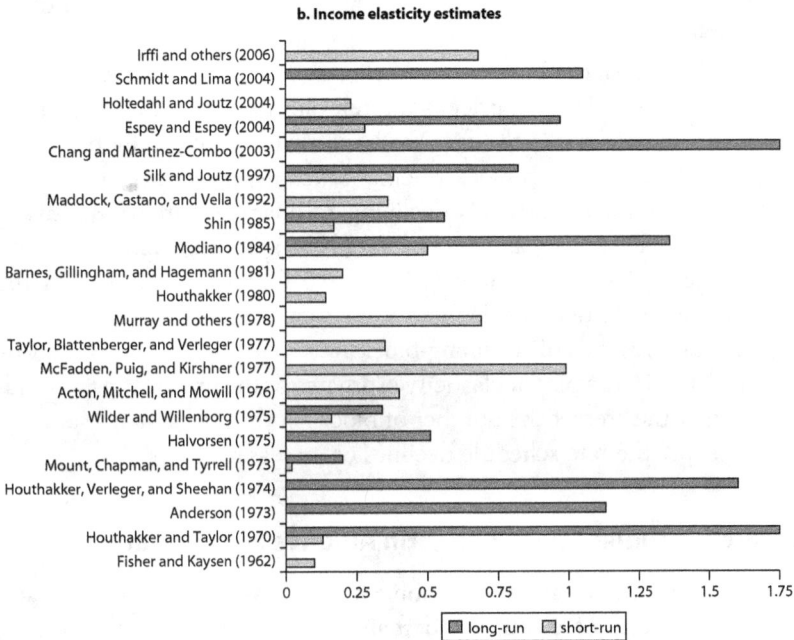

Moreover, as shown by studies such as Kamerschen and Porter (2004), there is a difference in magnitude between the residential and the industrial price and income elasticities of electricity demand. Estimates show that residential customers are more price-sensitive than industrial customers, and industrial variable price elasticities fluctuate less than do the residential estimates, which is consistent with the view that households spend a larger share of their budget on electricity. Likewise, a number of other studies, such as Barnes, Gillingham, and Hagemann (1981), who relate the households' level of electricity consumption to their stock of electrical appliances, find substantial variation in the residential electricity demand short-run price and income elasticities across *end-use categories*. Namely, although relatively higher price elasticity is common to the heating and air conditioning categories, water heating and lighting are generally price inelastic and less susceptible to marginal use changes despite their high average and potential usage level.

Finally, marked differences in opinion among academic researchers exist also with regard to the approach to measuring price and income elasticity of demand—specifically, whether the price and income elasticities should best be estimated at the *marginal* or the *average* price of electricity. For instance, Wilder and Willenborg (1975) defend the use of average price on the ground that the consumer responds to his total monthly bill and rarely knows his marginal rate. However, several other researchers, such as Houthakker, Verleger, and Sheehan (1974), Taylor (1975), Taylor, Blattenberger, and Verleger (1977), and Berndt (1984), have argued that marginal price is the relevant price variable. Part of the reason for this debate is the fact that electricity has typically been sold according to a "multistep block pricing" schedule, under which marginal price is a step function, usually declining, of quantity purchased. Studies by Hausman, Kinnucan, and McFadden (1979), Barnes, Gillingham, and Hagemann (1981), and Dubin (1982) have found the estimate of the marginal price elasticity of demand to be significantly biased away from 0.00 in the presence of declining-block rate schedules, and as stated by Henson (1984), the bias in elasticity estimates of the marginal price tends to be larger the greater the number of blocks in the rate schedule and the more steeply the rate schedule declines or increases.

Elasticity Studies Specific to Latin America and the Caribbean

In the Latin America and the Caribbean region, studies on the price and income elasticity of electricity demand are still relatively scarce and, similar to the research focusing on other parts of the world, display some variation in the specific estimated coefficients. However, most of the region-specific elasticity estimates tend to fall in the lower range of the overall spectrum. Estimates for Brazil carried out by Eletrobrás and researchers such as Schmidt and Lima (2004), for instance, show an income elasticity of demand at above unity for both the residential and the industrial sectors, illustrated, between 1980 and 2000, by the high average annual increase in electricity demand compared to the average annual growth in gross domestic product—at 5.4 percent and 2.4 percent, respectively. However, the long-run price elasticity is estimated at a very low level: −0.15 for the residential sector and −0.13 for the industrial sector. Irffi and others (2006), focusing on the country's Northeast and covering the period 1970–2003, specifically focus on the short-run income elasticity for residential demand, estimating it at about 0.84. Yet other researchers, such as Andrade and Lobão (1997), analyzing data for the period from 1963 to 1995, have found the short-run price and income elasticities to exceed the long-run ones. Finally, the models developed by Carlos, Notini, and Maciel (2009) not only highlight the comparatively higher sensitivity of residential consumers relative to industrial consumers to variations in the price of electricity, but also find that the exact elasticity coefficients need not be constant over time.

Studies on Mexico, such as Chang and Martinez-Chombo (2003), covering the 1985–2000 timeframe, similarly, find the long-run price elasticity of residential electricity demand to be fairly low—at about –0.44—with the industrial sector demand being even less elastic—at –0.25. With regard to demand responsiveness to changes in income, the studies estimate the long-run elasticities of the residential and the industrial sectors at 1.95 and 1.29, respectively. For Chile, Benavente, Galetovic, Sanhueza, and Serra (2005) estimate the price elasticity of electricity demand by commercial and residential users, finding the short-run values at –0.33 and –0.19 for residential and commercial consumers, respectively, while the long-term values are estimated at –0.41 and –0.21, respectively. Finally, the research on the topic focusing on Colombia, such as the study by Maddock, Castano, and Vella (1992), estimates the short-run price and income elasticities of residential electricity demand in the range between –0.17 and –0.47 and between 0.30 and 0.36, respectively. Interestingly, the authors also find a consistent pattern whereby higher-income consumers have absolutely larger price and income elasticities than do the poor ones.

Note

1. The highest responsiveness (as low as –0.27) to changes in the price of electricity was found in the water supply industry, while the steel tube industry was found to have the lowest price-elasticity value (–0.007).

References

Acton, J., B. Mitchell, and R. Mowill. 1976. "Residential Demand for Electricity in Los Angeles: An Econometric Study of Disaggregated Data." Report R-1899-NSF, The Rand Corporation, Santa Monica, CA.

Anderson, K. P. 1973. "Residential Demand for Electricity: Econometric Estimates for California and the United States." *Journal of Business* 46 (4): 526–53.

Andrade, T., and W. Lobão. 1997. "Elesticidade-renda e preço da demanda residencial de energía elétrica no Brasil." Texto para Discussão n 489, Instituto de Pesquisa Econômica Aplicada, Rio de Janeiro.

Barnes, R., R. Gillingham, and R. Hagemann. 1981. "The Short-Run Residential Demand for Electricity." *Review of Econometrics and Statistics* 63 (4): 541–51.

Beenstock, M., E. Goldin, and D. Nabot. 1999. "The Demand for Electricity in Israel." *Energy Economics* 21 (2): 168–83.

Benavente, J. M., A. Galetovic, R. Sanhueza, and P. Serra. 2005. "Estimando la Demanda Residencial por Electricidad en Chile: El Consumo es Sensible al Precio." *Cuadernos de Economía* 42 (May): 31–61.

Berndt, E. R. 1984. "Modeling the Aggregate Demand for Electricity: Simplicity versus Virtuosity." In *Advances in the Economics of Energy and Resources*, ed. J. R. Moroney, vol. 5, 141–52. Greenwich, CT: JAI Press.

Bjørner, T. B., and H. H. Jensen. 2002. "Interfuel Substitution within Industrial Companies: An Analysis Based on Panel Data at Company Level." *Energy Journal* 23 (2): 27–50.

Bohi, D. R., and M. Zimmerman. 1984. "An Update on Econometric Studies of Energy Demand." *Annual Review of Energy* 9: 105–54.

Boisvert, R. N., P. A. Cappers, and B. Neenan. 2002. "The Benefits of Customer Participation in Wholesale Electricity Markets." *Electric Journal* 15 (3): 41–51.

Boonekamp, P. G. M. 2007. "Price Elasticities, Policy Measures, and Actual Developments in Household Energy Consumption—A Bottom Up Analysis for the Netherlands." *Energy Economics* 29 (2): 133–57.

Branch, E. R. 1993. "Short-Run Income Elasticity of Demand for Residential Electricity Using Consumer Expenditure Survey Data." *Energy Journal* 14 (4): 111–22.

Brännlund, R., T. Ghalwash, and J. Nordström. 2007. "Increased Energy Efficiency and the Rebound Effect: Effects on Consumption and Emissions." *Energy Economics* 29 (1): 1–17.

Caloghirou, Y. D., A. G. Mourelatos, and H. Thompson. 1997. "Industrial Energy Substitution during the 1980s in the Greek Economy." *Energy Economics* 19 (4): 476–91.

Carlos, A. P., H. Notini, and L. F. Maciel. 2009. "Brazilian Electricity Demand Estimation: What Has Changed after the Rationing in 2001? An Application of Time Varying Parameter Error Correction Model." Graduate School of Economics, Getulio Vargas Foundation, Rio de Janiero.

Caves, D., K. Eakin, and A. Faruqui. 2000. "Mitigating Price Spikes in Wholesale Markets through Market-Based Pricing in Retail Markets." *The Electricity Journal* 13 (3): 13–23.

Chang, Y., and E. Martinez-Chombo. 2003. "Electricity Demand Analysis Using Cointegrating and Error-Correction Models with Time Varying Parameters: The Mexican Case." Working Paper 2003-08, Department of Economics, Rice University, Houston, TX.

Dubin, J. A. 1982. "Economic Theory and Estimation of the Demand for Consumer Durable Goods and Their Utilization: Appliance Choice and the

Demand for Electricity." Working Paper MIT-EL 82-035WP, MIT Energy Laboratory, Massachusetts Institute of Technology, Cambridge, MA.

Elkhafif, M. A. T. 1992. "Estimating Disaggregated Price Elasticities in Industrial Energy Demand." *Energy Journal* 13 (4): 209–17.

Energy Information Administration. 1990. "PC-AEO Forecasting Model for the Annual Energy Outlook 1990—Model Documentation." DOE/EIA-MO36 (90), U.S. Department of Energy, Washington, DC.

Espey, J., and M. Espey. 2004. "Turning on the Lights: A Meta-Analysis of Residential Electricity Demand Elasticities." *Journal of Agricultural and Applied Economics* 36 (1): 65–81.

Faruqui, A., and S. S. George. 2002. "The Value of Dynamic Pricing in Mass Markets." *Electricity Journal* 15 (6): 45–55.

Filippini, M., and S. Pachuari. 2002. "Elasticities of Electricity Demand in Urban Indian Households." CEPE Working Paper 16, Centre for Energy Policy and Economics, Swiss Federal Institutes of Technology, Zurich.

Fisher, F. M., and C. Kaysen. 1962. *A Study in Econometrics: The Demand for Electricity in the United States.* Amsterdam: North-Holland.

Galetovic, A., R. Sanhueza, and P. Serra. 2001. Estimacion de los costos de falla residencial y comerical. Draft. Santiago, Chile.

Halvorsen, R. 1975. "Residential Demand for Electric Energy." *Review of Economics and Statistics* 57 (1): 12–18.

Hausman, J. A., M. Kinnucan, and D. McFadden. 1979. "A Two-Level Electricity Demand Model: Evaluation of the Connecticut Time-of-Day Pricing Test." *Journal of Econometrics* 10 (3): 263–89.

Heffner, G. C., and C. A. Goldman. 2001. "Demand Responsive Programs—An Emerging Resource for Competitive Electricity Markets?" Report LBNL-48 374, Lawrence Berkeley National Laboratory, University of California-Berkeley.

Henson, S. 1984. "Electricity Demand Estimates under Increasing-Block Rates." *Southern Economic Journal* 51 (1): 147–56.

Hesse, D. M., and H. Tarkka. 1986. "The Demand for Capital, Labor and Energy in European Manufacturing Industry before and after the Oil Price Shocks." *Scandinavian Journal of Economics* 88.

Holtedahl, P., and F. L. Joutz. 2004. "Residential Electricity Demand in Taiwan." *Energy Economics* 26 (2): 201–24.

Houthakker, H. 1980. "Residential Electricity Revisited." *Energy Journal* 1 (1): 29–42.

Houthakker, H. S., and L. D. Taylor. 1970. *Consumer Demand in the United States,* 2nd ed. Cambridge, MA: Harvard University Press.

Houthakker, H. S., P. K. Verleger, and D. P. Sheehan. 1974. "Dynamic Demand Analyses for Gasoline and Residential Electricity." *American Journal of Agricultural Economics* 56 (2): 412–18.

Ilmakunnas, P., and H. Törmä. 1989. "Structural Change in Factor Substitution in Finnish Manufacturing." *Scandinavian Journal of Economics* 91 (4): 705–21.

Irffi, G., I. Castelar, M. Siqueira, and F. Linhares. 2006. "Dynamic OLS and Regime Switching Models to Forecast the Demand for Electricity in the Northeast of Brazil." Graduate School of Economics, Getulio Vargas Foundation. Rio de Janie.

Jones, C. T. 1995. "A Dynamic Analysis of Interfuel Substitution in U.S. Industrial Energy Demand." *Journal of Business and Economic Statistics* 13 (4): 459–65.

Kamerschen, D. R., and D. V. Porter. 2004. "The Demand for Residential, Industrial, and Total Electricity, 1973–1998." *Energy Economics* 26 (1): 87–100.

Kirschen, D. S. 2003. "Demand-Side View of Electricity Markets." *IEEE Transactions on Power Systems* 18 (2): 520–27.

Lijesen, M. G. 2007. "The Real-Time Price Elasticity of Electricity." *Energy Economics* 29 (2): 249–58.

Lyman, R. A. 1973. "Price Elasticities in the Electric Power Industry." Department of Economics, University of Arizona, Tucson.

Maddock, R., E. Castano, and F. Vella. 1992. "Estimating Electricity Demand: The Cost of Linearising the Budget Constraint." *Review of Economics and Statistics* 74 (2): 350–54.

McFadden, D., C. Puig, and D. Kirshner. 1977. "A Simulation Model for Electricity Demand." Final Report, Cambridge Systematics, Inc., Cambridge, MA.

Modiano. E. M. 1984. "Elasticidade-renda e preço da demanda de energia elétrica no Brasil." Rio de Janeiro: PUC/RJ (Texto para discussão, 68).

Mount, T., L. Chapman, and T. Tyrrell. 1973. "Electricity Demand in the United States: An Econometric Analysis." ORNL-NSF-EP-49, Oak Ridge National Laboratories, Oak Ridge, Tennessee.

Murray, M., R. Spann, L. Pulley, and E. Beauvais. 1978. "The Demand for Electricity in Virginia." *Review of Economics and Statistics* 60 (4): 585–600.

Roos, J. G., and I. E. Lane. 1998. "Industrial Power Demand Response Analysis for One-Part Real-Time Pricing." *IEEE Transactions on Power Systems* 13 (1): 159–64.

Roy, J., A. H. Sanstad, J. A. Sathaye, and R. Khaddaria. 2006. "Substitution and Price Elasticity Estimates Using Inter-Country Pooled Data in a Translog Cost Model." *Energy Economics* 28 (5–6): 706–19.

Schmidt, C. A., and M. A. M. Lima. 2004. "A demanda por energía elétrica no Brasil." *Revista Brasileira de Economia* 58 (1): 67–98.

Shin, J.-S. 1985. "Perception of Price When Price Information Is Costly: Evidence from Residential Electricity Demand." *Review of Economics and Statistics* 67 (4): 591–98.

Silk, J. I., and F. L. Joutz. 1997. "Short and Long-Run Elasticities in U.S. Residential Electricity Demand: A Co-integration Approach." *Energy Economics* 19 (4): 493–513.

Taheri, A. A. 1994. "Oil Shocks and the Dynamics of Substitution Adjustments of Industrial Fuels in the US." *Applied Economics* 26 (1994) (8).

Taylor, L. D. 1975. "The Demand for Electricity: A Survey." *Bell Journal of Economics* 6 (1): 74–110.

Taylor, L. D., G. R. Blattenberger, and P. K. Verleger Jr. 1977. "The Residential Demand for Energy." Final Report, EPRI EA-235, vol. 1, Electric Power Research Institute, Palo Alto, CA.

Urga, G., and C. Walters. 2003. "Dynamic Translog and Linear Logit Models: A Factor Demand Analysis of Interfuel Substitution in U.S. Industrial Energy Demand." *Energy Economics* 25 (1): 1–21.

Wilder, R. P., and J. F. Willenborg. 1975. "Residential Demand for Electricity: A Consumer Panel Approach." *Southern Economic Journal* 42 (2): 212–17.

Wilson, J. W. 1971. "Residential Demand for Electricity." *Quarterly Review of Economics and Business* 11 (1): 7–22.

Wolak, F. A., and R. H. Patrick. 2001. "Estimating the Customer-Level Demand for Electricity under Real Time Market Prices." NBER Working Paper 8213, National Bureau of Economic Research, Cambridge, MA.

Woodland, A. D. 1993. "A Micro-econometric Analysis of the Industrial Demand for Energy in NSW." *The Energy Journal* 14 (1993) (2), 57–89.

Yusta, J. M., and J. A. Dominguez. 2002. "Measuring and Modeling of Industrial Demand Response to Alternative Prices of the Electricity." 14th Power Systems Computation Conference, Session 15, Paper 3, Seville, Spain, June 24–28.

Zachariadis, T., and N. Pashourtidou. 2007. "An Empirical Analysis of Electricity Consumption in Cyprus." *Energy Economics* 29 (2): 183–98.

APPENDIX C

Electricity Supply Model and Fuel Price Assumptions

Description of Electricity Supply Model

After the total annual demand is calculated, the SUPER (Sistema Unificado de Planificación Eléctrica Regional) model by OLADE (Organización Latinoamericana de Energía, or Latin American Energy Organization) is then used to determine the optimal, cost-minimizing generation mix to meet the demand. The SUPER model was developed by OLADE and is aimed at the prioritization, scaling, and selection of electricity projects to meet the growth in electricity demand. In each phase, the system determines generation targets for each of the system's power plants, minimizes the expected value of the operating and capital costs throughout the period, and evaluates the financial and environmental effects caused by the future development of the electricity sector.

In addition to the demand scenario as an input to the model, various data are also inputted into the SUPER model, including hydrology; reference prices for fuels; existing plants with their operational features; projects under construction or in the bidding process, which are fixed, and their entry dates; and eligible projects with their earlier entry dates and operational features, investment costs, and operational variables, among other inputs.

The modeling of demand curves is done on the basis of historical demand records, which form the starting point for the load curves. Thus, according to annual estimates of the long-term demand scenario obtained from the detailed analysis performed using the electricity demand model and using historical demand records, the SUPER model generates energy and power demands of each year of the prospective horizon.

The model uses country-specific information on hydrology, when available. This information includes the time series of volumes of flow in different sites, which preserves the most important time and space parameters estimated on the basis of historical records. Its objective is to supply hydrological information for optimization and simulation. The model produces the available and minimal energy, maximum capacity, and storable energy for each hydroelectric project, period of time, and hydrological condition.

For production of these results and use of SUPER, hydrological data for different measurement stations must be inputted, and these data must be related to each basin with operational or eligible projects in that basin.

The problem of expansion planning can be divided into two issues: investment and operation. These problems are due to the nature of the two-stage decision-making process of first making expansion decisions, and then evaluating those decisions once in operation. Using operating costs and other factors, one can reformulate the expansion strategy, which in turn affects the operations side once again.

The analysis of system expansion under the SUPER model uses a path of capacity expansion, with selection of hydroelectric projects and the expansion of the so-called thermal classes (grouping of thermal plants with similar technological and operational features, such as steam plants that use coal). This expansion path is obtained by minimizing the total costs of investment and operation throughout the study period. The expansion plans for hydroelectric capacity consider the country expansion plans.

As a result, the capacity expansion plan is obtained, with the addition of individual hydroelectric projects and of thermoelectric projects grouped in thermal classes, with the installed capacities in each case and the start-up dates. Using these data, one calculates a power balance to verify the coverage of the system's power demand or maximum demand and the gross margin of reserve. The investments required for this plan are also calculated.

The expected generation for each period of each hydroelectric plant and of each thermal class is obtained from the model. The annual generation by plant and type of plant and the system's energy balance must be calculated separately using text files transferred to Microsoft Excel. In addition, the analysis and verification of results are conducted separately from the SUPER model.

Supply Assumptions

Plant Specification Assumptions by Technology

Given the current available information on costs, plant dependability, and state of development, the model considers the technologies shown in table C.1. To simplify the assumptions, one assumes that plant specifications are constant across the region. The unitary investment cost for each technology except hydroelectricity-based generation is constant for every country.[1] Other technologies, including solar power (both concentrated solar thermal and photovoltaic), were not considered in this analysis.

Carbon Tax Assumptions

Two carbon tax scenarios were used for US$20 and US$50 per ton of carbon dioxide (CO_2), respectively. The assumed additional cost per fuel type is shown in table C.2.

Table C.1 Levelized Costs by Technology

	Levelized costs, US$ per MWh		
Crude price	50	100	150
Combined-cycle (natural gas)	44	64	78
Hydropower	39	39	39
Coal	41	49	52
Geothermal energy	77	77	77
Nuclear energy	78	78	78
Wind	93	93	93
Diesel and fuel oil	140	207	273

Source: Authors' elaboration.
Note: MWh = megawatt hour.

Table C.2 Additional Costs per Fuel Type under Carbon Tax Scenario

	Diesel (US$/Barrel)	Fuel oil (US$/Barrel)	Natural gas (US$/MMBtu)	Coal (US$/Ton)
20 US$/Ton CO_2	7.95	9.9	1.0	78.36
50 US$/Ton CO_2	19.87	24.7	2.4	195.90

Source: Authors' elaboration.
Note: MMBtu = million British thermal units.

Fuel Price Assumptions

The crude oil price is a key parameter for the cost of electricity across the generation technologies. For production of the oil price sensitivity analysis, it is important to define the associated prices for natural gas, coal, and oil derivatives.

For the price of fuel oil at the plant door, the model assumes transport and handling costs for only coal and natural gas (related to pipeline transport cost). Accordingly, the final cost of these fuels is computed as the sum of the spot price and the transport and handling cost. However, the cost for diesel and heavy fuel oil is computed only as the spot price. It is thus worth emphasizing an implicit assumption made by the model: there is no cost to use wind and water.[2] The spot prices for natural gas, fuel oil, diesel, and coal used throughout this report for the 20-year time analysis are based on the projections provided with the assumption of fixed prices for each fuel along the study period. We considered a price for oil of US$100 per barrel. To simplify our exercise and the assumptions around fuel prices, we estimate a price for coal, natural gas, and oil derivatives considering the equations below.[3] These equations result from least square regressions of the price of each fuel against the oil price (as an average of crude oil WTI [West Texas intermediate]). The equations used are as follows[4]:

Oil price: $OilP_t$
Coal price: $CP_t = 30.4 + 0.611 \times OilP_t$
Natural gas price: $GasP_t = 2.5 + 0.058 \times OilP_t$
Heavy fuel oil price: $HFOP_t = 0.4 + 0.818 \times OilP_t$
Diesel price: $DP_t = 0.57 + 1.194 \times OilP_t$

Notes

1. For hydro-based generation, the investment cost assumptions for Central America and for Colombia, Ecuador, and Peru were a specific unitary cost for each project based on the information contained in each country's expansion plan. For other countries, the average unitary cost was assumed to be a function of the marginal availability of hydrologic resources. Hence, the average unitary cost of installed capacity for Brazil was US$1,800 per kilowatt (kW), for Mexico and Paraguay US$2,500 per kW, and for all other countries US$2,000 per kW.

2. It must be stressed that this assumption of zero cost for water is just an average premise for all the countries of Latin American and the Caribbean.

3. The study team is fully aware of the simplification of this analysis. Prices for oil, natural gas, and coal are not fully correlated. Furthermore, the economics explaining prices for these fuels consider factors such as demand, relative scarcity, and substitutability among other factors. The fundamentals of these fuel markets are quite complex, and a deeper analysis of the future tendencies of the prices of these fuels is out of the scope of this report. Hence, we simplify the exercise by considering a simple regression analysis approach.

4. The equations refer to the coal price in US$ per ton, natural gas price in US$ per MMBtu (million British thermal units), heavy fuel oil price in US$ per barrel, diesel price in US$ per barrel, and crude oil price in US$ per barrel.

Comparison of ICEPAC Results with Country Expansion Plans

	Comparison document	ICEPAC
Argentina		
Document	Balance Energético Nacional (Presentation: La Política Energética Argentina: Elementos para el Planeamiento Energético)	ICEPAC Scenario
Publication date	2008	2010
Period covered	2008–25	2008–30
Electricity demand	216.4 TWh (2025)	168.0 TWh (2025)
Electricity demand (annual growth, %)	3.2 (average annual growth, 2008–25)	2.7 (average annual growth, 2008–30)
GDP growth assumptions (%)	2008–14: 4.0 2014–18: 3.0 2018–25: 2.5	2.9 (average annual growth, 2008–30)
Bolivia		
Document	Actualización del Plan Referencial del Sistema Interconectado Nacional Boliviano	ICEPAC Scenario
Publication date	2005	2010
Period covered	2005–14	2008–30
Electricity demand	6.136 TWh (2014)	9.030 TWh (2014)
Electricity demand (annual growth, %)	5.2 (average annual growth, 2005–14)	6.5 (average annual growth, 2008–30)
GDP growth assumptions (%)	4.0 average annual GDP growth 2005–14 in the base case; pessimistic case: 2.6; optimistic case: 5.4	3.2 (average annual growth, 2008–30)

Brazil

		ICEPAC Scenario
Document	"Plano Nacional de Energia 2030." Ministério de Minas e Energia	
Publication date	2007	2010
Period covered	2005–30	2008–30
Electricity demand	859–1,245 TWh (range of values within four growth scenarios, 2030)	1,087 TWh (2030)
Electricity demand (annual growth, %)	3.5–5.1 (range of values within four growth scenarios, 2005–30)	4.4 (average annual growth, 2008–30)
GDP growth assumptions (%)	2.2–5.1 (range of values within four growth scenarios, 2005–30)	3.0 (average annual growth, 2008–30)

Colombia

		ICEPAC Scenario
Document	"Plan de Expansión de Referencia Generación—Transmisión 2009–2023." República de Colombia, Ministerio de Minas y Energía	
Publication date	2008	2010
Period covered	2009–25	2008–30
Electricity demand	89.0–111.6 TWh (2025)	82.1 TWh (2025)
Electricity demand (annual growth, %)	3.3 (average annual growth, 2009–25)	3.1 (average annual growth, 2008–30)
GDP growth assumptions (%)	6.0 (sustained annual GDP growth, 2009–25)	3.1 (average annual growth, 2008–30)

(continued next page)

Costa Rica

		ICEPAC Scenario
Document	"Plan de Expansión de la Generación Eléctrica, Periodo 2008–2021." Instituto Costarricense de Electricidad	
Publication date	2007	2010
Period covered	2008–21 (although demand projections go through 2030)	2008–30
Electricity demand	Pessimistic case: 21.2 TWh; optimistic case: 24.9 TWh (2030)	21.8 TWh (2030)
Electricity demand (annual growth, %)	3.2–4.6 (average annual growth, 2008–30)	3.4 (average annual growth, 2008–30)
GDP growth assumptions (%)	3.0 (average annual growth, 2008–30)	3.1 (average annual growth, 2008–30)

Ecuador

		ICEPAC Scenario
Document	"Plan Maestro de Electrificación del Ecuador 2007–2016." CONELEC (Consejo Nacional de Electricidad)	
Publication date	2007	2010
Period covered	2007–16	2008–30
Electricity demand	Base case: 22.82 TWh (2017)	22.10 TWh (2017)
Electricity demand (annual growth, %)	3.56 (average annual growth, 2008–17)	4.80 (average annual growth, 2008–30)
GDP growth assumptions (%)	4.72 (average annual growth, 2008–17)	2.70 (average annual growth, 2008–30)

El Salvador

		ICEPAC Scenario
Document	"National Expansion Plan"	
Publication date	2003	2010
Period covered	2003–20	2008–30

Electricity demand	9.38 TWh (2020)	8.50 TWh (2020)
Electricity demand (annual growth, %)	4.7 (average annual growth, 2003–20)	4.1 (average annual growth, 2008–30)
GDP growth assumptions (%)	4.8 (average annual growth, 2003–20)	2.8 (average annual growth, 2008–30)

Guatemala

Document	"Plan de Expansión Indicativo del Sistema de Generación." Comisión Nacional de Energía Eléctrica	ICEPAC Scenario
Publication date	2008	2010
Period covered	2008–22	2008–30
Electricity demand	13.4–19.3 TWh (range of values within four scenarios for energy as a whole, 2008–22)	17.3 TWh (2022)
Electricity demand (annual growth, %)	4.6–6.8 (range of values within four different scenarios for energy as a whole, 2008–22)	6.6 (average annual growth, 2008–30)
GDP growth assumptions (%)	Pessimistic case: 3.3; optimistic case: 5.5	3.1 (average annual growth, 2008–30)

Honduras

Document	"Proyección de Demanda de Energía Eléctrica, Escenario Base." Empresa Nacional de Energía Eléctrica	ICEPAC Scenario
Publication date	2010	2010
Period covered	2010–25	2008–30
Electricity demand	14.47 TWh (2025)	11.50 TWh (2025)
Electricity demand (annual growth, %)	6.41 (average annual growth, 2010–25)	5.30 (average annual growth, 2010–25)
GDP growth assumptions (%)	—	2.9 (average annual growth, 2008–30)

(continued next page)

Mexico

Document	"Prospectiva del Sector Eléctrico." Secretaría de Energía	ICEPAC Scenario
Publication date	2009	2010
Period covered	2009–24	2008–30
Electricity demand	365 TWh (2024)	373 TWh (2024)
Electricity demand (annual growth, %)	3.2 (average annual growth, 2009–24)	3.2 (average annual growth, 2008–30)
GDP growth assumptions (%)	2.7 (average annual growth, 2009–24)	2.8 (average annual growth, 2008–30)

Nicaragua

Document	"Plan Indicativo de Generación 2008–2014 (Escenarios Referenciales)." Ministerio de Energía y Minas	ICEPAC Scenario
Publication date	2008	2010
Period covered	2008–14	2008–30
Electricity demand	4.2 TWh (2014)	4.3 TWh (2014)
Electricity demand (annual growth, %)	5.0 (average annual growth, 2008–14)	5.0 (average annual growth, 2008–30)
GDP growth assumptions (%)	3.6 (average annual growth, 2008–14)	2.8 (average annual growth, 2008–30)

Panama

Document	"Plan de Expansión del Sistema Interconectado Nacional 2007–2021." Empresa de Transmisión Eléctrica S.A.	ICEPAC Scenario
Publication date	2007	2010
Period covered	2007–21	2008–30

Electricity demand	12.29 TWh (2021)	13.60 TWh (2021)
Electricity demand (annual growth, %)	4.4 (average annual growth, 2008–30)	4.6 (average annual growth, 2008–30)
GDP growth assumptions (%)	2.7 (average annual growth, 2008–30)	3.5 (average annual growth, 2008–30)

Peru

Document	"Peru Plan Referencial de Electricidad" Ministerio de Energía y Minas	ICEPAC Scenario
Publication date	2006	2010
Period covered	2006–15	2008–30
Electricity demand	Base case: 43.74 TWh (2015)	38.50 TWh (2015)
Electricity demand (annual growth, %)	Base case: 6.6 (average annual growth, 2006–15)	4.2 (average annual growth, 2008–30)
GDP growth assumptions (%)	Base case: 6.5 (average annual growth, 2006–15)	3.9 (average annual growth, 2008–30)

Source: Authors' compilation.

Note: GDP = gross domestic product, GWh = gigawatt-hour, ICEPAC = Illustrative Country Expansion Plans Adjusted and Constrained, TWh = terawatt-hour, — = not available.

Bibliography

Acton, J., B. Mitchell, and R. Mowill. 1976. "Residential Demand for Electricity in Los Angeles: An Econometric Study of Disaggregated Data." Report R-1899-NSF, The Rand Corporation, Santa Monica, CA.

Al Faris, A. R. F. 2002. "The Demand for Electricity in the GCC Countries," *Energy Policy* 30 (2): 117–24.

Anderson, K. P. 1973. "Residential Demand for Electricity: Econometric Estimates for California and the United States." *Journal of Business* 46 (4): 526–53.

Andrade, T., and W. Lobão. 1997. "Elesticidade-renda e preço da demanda residencial de energía elétrica no Brasil." Texto para Discussão n 489, Instituto de Pesquisa Econômica Aplicada, Rio de Janeiro.

Barnes, R., R. Gillingham, and R. Hagemann. 1981. "The Short-Run Residential Demand for Electricity." *Review of Econometrics and Statistics* 63 (4): 541–51.

Beenstock, M., E. Goldin, and D. Nabot. 1999. "The Demand for Electricity in Israel." *Energy Economics* 21 (2): 168–83.

Berndt, E. R. 1984. "Modeling the Aggregate Demand for Electricity: Simplicity versus Virtuosity." In *Advances in the Economics of Energy and Resources*, ed. J. R. Moroney, vol. 5, 141–52. Greenwich, CT: JAI Press.

Bjørner, T. B., and H. H. Jensen. 2002. "Interfuel Substitution within Industrial Companies: An Analysis Based on Panel Data at Company Level." *Energy Journal* 23 (2): 27–50.

Bohi, D. R. 1981. *Analyzing Demand Behavior: A Study of Energy Elasticities.* Washington, DC: RFF Press.

———. 1984. *Price Elasticities of Demand for Energy: Evaluating the Estimates.* Washington, DC: RFF Press.

Bohi, D. R., and M. Zimmerman. 1984. "An Update on Econometric Studies of Energy Demand." *Annual Review of Energy* 9: 105–54.

Boisvert, R., P. Cappers, B. Neenan, and B. Scott. 2004. *Industrial and Commercial Customer Response to Real Time Electricity Prices.* Syracuse, NY: Neenan Associates.

Boisvert, R. N., P. A. Cappers, and B. Neenan. 2002. "The Benefits of Customer Participation in Wholesale Electricity Markets." *Electric Journal* 15 (3): 41–51.

Boonekamp, P. G. M. 2007. "Price Elasticities, Policy Measures and Actual Developments in Household Energy Consumption—A Bottom Up Analysis for the Netherlands." *Energy Economics* 29 (2): 133–57.

Branch, E. R. 1993. "Short-Run Income Elasticity of Demand for Residential Electricity Using Consumer Expenditure Survey Data." *Energy Journal* 14 (4): 111–22.

Brännlund, R., T. Ghalwash, and J. Nordström. 2007. "Increased Energy Efficiency and the Rebound Effect: Effects on Consumption and Emissions." *Energy Economics* 29 (1): 1–17.

Caloghirou, Y. D., A. G. Mourelatos, and H. Thompson. 1997. "Industrial Energy Substitution during the 1980s in the Greek Economy." *Energy Economics* 19 (4): 476–91.

Carlos, A. P., H. Notini, and L. F. Maciel. 2009. "Brazilian Electricity Demand Estimation: What Has Changed after the Rationing in 2001? An Application of Time Varying Parameter Error Correction Model." May 5.

Caves, D., K. Eakin, and A. Faruqui. 2000. "Mitigating Price Spikes in Wholesale Markets through Market-Based Pricing in Retail Markets." *The Electricity Journal* 13 (3): 13–23.

Dahl, C. 1993. "A Survey of Energy Demand Elasticities in Support of the Development of the NEMS." Report prepared for the U.S. Department of Energy, Colorado School of Mines, Golden, CO.

Dubin, J. A. 1982. "Economic Theory and Estimation of the Demand for Consumer Durable Goods and Their Utilization: Appliance Choice and the Demand for Electricity." Working Paper MIT-EL 82-035WP, MIT Energy Laboratory, Massachusetts Institute of Technology, Cambridge, MA.

Elkhafif, M. A. T. 1992. "Estimating Disaggregated Price Elasticities in Industrial Energy Demand." *Energy Journal* 13 (4): 209–17.

Energy Information Administration. 1990. "PC-AEO Forecasting Model for the Annual Energy Outlook 1990—Model Documentation." DOE/EIA-MO36 (90), U.S. Department of Energy, Washington, DC.

Espey, M. 1999. "Turning on the Lights: A Meta-Analysis of Residential Electricity Demand Analysis." Working Paper, University of Nevada at Reno.

Espey, J., and M. Espey. 2004. "Turning on the Lights: A Meta-Analysis of Residential Electricity Demand Elasticities." *Journal of Agricultural and Applied Economics* 36 (1): 65–81.

Faruqui, A., and S. S. George. 2002. "The Value of Dynamic Pricing in Mass Markets." *Electricity Journal* 15 (6): 45–55.

Filippini, M., and S. Pachuari. 2002. "Elasticities of Electricity Demand in Urban Indian Households." CEPE Working Paper 16, Centre for Energy Policy and Economics, Swiss Federal Institutes of Technology, Zurich.

Fisher, F. M., and C. Kaysen. 1962. *A Study in Econometrics: The Demand for Electricity in the United States.* Amsterdam: North-Holland.

Halvorsen, R. 1973. "Demand for Electric Power in the United States." Discussion Paper 73-13, Institute for Economic Research, University of Washington, Seattle.

———. 1975. "Residential Demand for Electric Energy." *Review of Economics and Statistics* 57 (1): 12–18.

Hausman, J. A., M. Kinnucan, and D. McFadden. 1979. "A Two-Level Electricity Demand Model: Evaluation of the Connecticut Time-of-Day Pricing Test." *Journal of Econometrics* 10 (3): 263–89.

Heffner, G. C., and C. A. Goldman. 2001. "Demand Responsive Programs—An Emerging Resource for Competitive Electricity Markets?" Report LBNL-48 374, Lawrence Berkeley National Laboratory, University of California-Berkeley.

Henson, S. 1984. "Electricity Demand Estimates under Increasing-Block Rates." *Southern Economic Journal* 51 (1): 147–56.

Holtedahl, P., and F. L. Joutz. 2004. "Residential Electricity Demand in Taiwan." *Energy Economics* 26 (2): 201–24.

Houthakker, H. 1980. "Residential Electricity Revisited." *Energy Journal* 1 (1): 29–42.

Houthakker, H. S., and L. D. Taylor. 1970. *Consumer Demand in the United States,* 2nd ed. Cambridge, MA: Harvard University Press.

Ilmakunnas, P., and H. Törmä. 1989. "Structural Change in Factor Substitution in Finnish Manufacturing." *Scandinavian Journal of Economics* 91 (4): 705–21.

Jones, C. T. 1995. "A Dynamic Analysis of Interfuel Substitution in U.S. Industrial Energy Demand." *Journal of Business and Economic Statistics* 13 (4): 459–65.

Kamerschen, D. R., and D. V. Porter. 2004. "The Demand for Residential, Industrial, and Total Electricity, 1973–1998." *Energy Economics* 26 (1): 87–100.

Kirschen, D. S. 2003. "Demand-Side View of Electricity Markets." *IEEE Transactions on Power Systems* 18 (2): 520–27.

Kirschen, D. S., G. Strbac, P. Cumperayot, and D. de Paiva Mendes. 2000. "Factoring the Elasticity of Demand in Electricity Prices." *IEEE Transactions on Power Systems* 15 (2): 612–17.

Lijesen, M. G. 2007. "The Real-Time Price Elasticity of Electricity." *Energy Economics* 29 (2): 249–58.

Lim, H. B. F., and G. P. Jenkins. 2000. "Electricity Demand and Electricity Value." Development Discussion Paper 2000-1, JDI Executive Program, Queen's University, Kingston, Ontario, Canada.

Lyman, R. A. 1973. "Price Elasticities in the Electric Power Industry." Department of Economics, University of Arizona, Tucson.

Maddock, R., E. Castano, and F. Vella. 1992. "Estimating Electricity Demand: The Cost of Linearising the Budget Constraint." *Review of Economics and Statistics* 74 (2): 350–54.

McFadden, D., C. Puig, and D. Kirshner. 1977. "A Simulation Model for Electricity Demand." Final Report, Cambridge Systematics, Inc., Cambridge, MA.

McMahon, J. E. 1987. "The LBL Residential Energy Model: An Improved Policy Analysis Tool." *Energy Systems and Policy* 10 (1): 47–71.

Murray, M., R. Spann, L. Pulley, and E. Beauvais. 1978. "The Demand for Electricity in Virginia." *Review of Economics and Statistics* 60 (4): 585–600.

Nelson, C. R., and S. C. Peck. 1985. "The NERC Fan: A Retrospective Analysis of the NERC Summary Forecasts." *Journal of Business Economic Statistics* 3 (3): 179–87.

Nelson, C. R., S. C. Peck, and R. G. Uhler. 1989. "The NERC Fan in Retrospect and Lessons for the Future." *Energy Journal* 10 (2): 91–107.

Nordin, J. 1976. "A Proposed Modification of Taylor's Demand Analysis: Comment." *Bell Journal of Economics* 7 (2): 719–21.

OECD (Organisation for Economic Co-operation and Development). 2005. *OECD Economic Surveys: Brazil.* Paris: OECD Publications.

Roos, J. G., and I. E. Lane. 1998. "Industrial Power Demand Response Analysis for One-Part Real-Time Pricing." *IEEE Transactions on Power Systems* 13 (1): 159–64.

Roy, J., A. H. Sanstad, J. A. Sathaye, and R. Khaddaria. 2006. "Substitution and Price Elasticity Estimates Using Inter-Country Pooled Data in a Translog Cost Model." *Energy Economics* 28 (5–6): 706–19.

Schmidt, C. A., and M. A. M. Lima. 2004. "A demanda por energía elétrica no Brasil." *Revista Brasileira de Economia* 58 (1): 67–98.

Shin, J.-S. 1985. "Perception of Price When Price Information Is Costly: Evidence from Residential Electricity Demand." *Review of Economics and Statistics* 67 (4): 591–98.

Silk, J. I., and F. L. Joutz. 1997. "Short and Long-Run Elasticities in U.S. Residential Electricity Demand: A Co-integration Approach." *Energy Economics* 19 (4): 493–513.

Smith, V. K. 1980. "Estimating the Price Elasticity of U.S. Electricity Demand." *Energy Economics* 2 (2): 81–85.

Taylor, L. D. 1975. "The Demand for Electricity: A Survey." *Bell Journal of Economics* 6 (1): 74–110.

Taylor, L. D., G. R. Blattenberger, and P. K. Verleger Jr. 1977. "The Residential Demand for Energy." Final Report, EPRI EA-235, vol. 1, Electric Power Research Institute, Palo Alto, CA.

Urga, G., and C. Walters. 2003. "Dynamic Translog and Linear Logit Models: A Factor Demand Analysis of Interfuel Substitution in U.S. Industrial Energy Demand." *Energy Economics* 25 (1): 1–21.

Wilder, R. P., and J. F. Willenborg. 1975. "Residential Demand for Electricity: A Consumer Panel Approach." *Southern Economic Journal* 42 (2): 212–17.

Wilson, J. W. 1971. "Residential Demand for Electricity." *Quarterly Review of Economics and Business* 11 (1): 7–22.

Wolak, F. A., and R. H. Patrick. 2001. "Estimating the Customer-Level Demand for Electricity under Real Time Market Prices." NBER Working Paper 8213, National Bureau of Economic Research, Cambridge, MA.

Yusta, J. M., and J. A. Dominguez. 2002. "Measuring and Modeling of Industrial Demand Response to Alternative Prices of the Electricity." 14th Power Systems Computation Conference, Session 15, Paper 3, Seville, Spain, June 24–28.

Zachariadis, T., and N. Pashourtidou. 2007. "An Empirical Analysis of Electricity Consumption in Cyprus." *Energy Economics* 29 (2): 183–98.

www.ingramcontent.com/pod-product-compliance
Lightning Source LLC
Chambersburg PA
CBHW070911270326
41927CB00011B/2533